Teaching
Chemistry

to KS4

Elaine **Wilson**

non-specialist **handbook**

Stoughton

HODDER HEADLINE GROUP

We are grateful to the following examining bodies for permission to reproduce examination questions: QCA, SEG and MEG/OCR. The answers given within the book are entirely the responsibility of the author and have not been provided or approved by the exam boards.

Orders: please contact Bookpoint Ltd, 78 Milton Park, Abingdon, Oxon OX14 4TD. Telephone: (44) 01235 827720, Fax: (44) 01235 400454. Lines are open from 9.00–6.00, Monday to Saturday, with a 24 hour message answering service. E-mail address: orders@bookpoint.co.uk

A catalogue record for this title is available from The British Library

ISBN 0 340 73764 6

First published 1999
Impression number   10 9 8 7 6 5 4 3 2 1
Year   2005 2004 2003 2002 2001 2000 1999

Cover photo from Oxford Scientific Films.

Typeset by Wearset, Boldon, Tyne and Wear.
Printed in Great Britain for Hodder and Stoughton Educational, a division of Hodder Headline Plc, 338 Euston Road, London NW1 3BH by Redwood Books, Trowbridge, Wilts.

Contents

Introduction   **1**

1

Classifying Materials   **5**

Chapter 1   Solids, liquids, gases and the particle theory   5
Chapter 2   Elements, compounds and mixtures   21
Chapter 3   Atomic structure and bonding   29

2

Changing Materials   **38**

Chapter 4   Geological changes   38
Chapter 5   Chemical reactions   49
Chapter 6   Electrolysis   60
Chapter 7   Equations and quantitative chemistry   72
Chapter 8   Useful products from oil   80
Chapter 9   Useful products from metals   94
Chapter 10   Useful products from air and changes to the atmosphere   106

3

Patterns of Behaviour   **115**

Chapter 11   The periodic table   115
Chapter 12   Acids and bases   121

Examination questions annotated with
specimen answers   **128**

Contact addresses   **155**

Index   **159**

iii

Introduction

'Joy may be inarticulate, but reflection is empty without understanding. There is delight to be had merely by looking at the world, but that delight can be deepened when the mind's eye can penetrate the surface of things to see the connections within.' P.W. Atkins

It is now usual that science teachers at Key Stage 3 will teach across the discipline and this pattern frequently continues into Key Stage 4. There is little evidence to confirm fears held by some science educators that pupils are dissuaded from studying the sciences beyond 16 when taught by a non-specialist. In fact it is more likely that the effective biology or physics teacher will try to be even more effective when teaching chemistry. The purpose of this book is to offer advice to the non-chemist on the problems children may encounter in learning chemistry, and to suggest tried and tested strategies which may help teach key chemistry concepts at Key Stages 3 and 4.

The excitement of using Bunsen burners and handling 'dangerous' chemicals has ensured that the early stages of secondary school chemistry are enjoyed by most pupils. Research evidence suggests, however, that by the end of year nine, many pupils consider chemistry to be boring and very difficult. There have been a number of suggestions as to why this is the case. It may be that pupils are unable to see the relevance of chemistry to their everyday lives. The Salters' courses and SATIS materials have been developed in an attempt to rectify this and many schools already use these resources. Others have suggested that the difficulties may be due to the nature of the subject and the way abstract chemical concepts are introduced to the young learner. Professor Alex Johnstone at Glasgow University proposes that modern chemistry has three basic components

1 The *macrochemistry* of the tangible, edible and visible

2 The *sub-microchemistry* of the molecular, atomic and kinetic

1

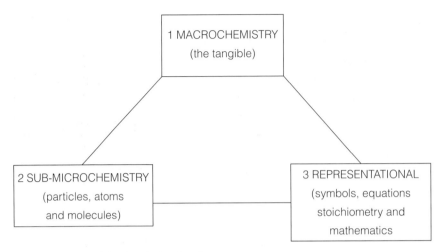

The three apices of the chemistry triangle

3 The *representational* chemistry of symbols, equations, stoichiometry and mathematics.

The professional chemist and chemistry teacher operate inside the triangle represented in the diagram, with a blend of macro, sub-micro and representational modes. They easily work within the triangle as the thinking requires. The difficulties articulated by pupils might be the result of their chemistry lessons operating within this triangle without prior preparation. As teachers of chemistry, it seems vital that we are aware of the potential difficulties some learners have with chemical concepts and that we consider these in our lesson planning.

There are many theories of how children learn, and I believe it is important that all teachers have a theory and that our practice is influenced by this. Losing sight of the purpose of teaching science and the problems of the learner can lead to rote learning and a 'coverage of the syllabus' mentality. Some of the dissatisfaction with chemistry lessons expressed by year nine pupils might be attributed to such a teaching approach. One model which might be useful to the teacher of chemistry, which acknowledges the difficulties suggested by Johnstone's triangle, uses an information processing theory of learning that has grown up alongside the development of the computer (see diagram on page 3).

Pupils are often presented with lots of information in chemistry lessons, along with visual stimuli, smells and bangs. The processing model of learning suggests that pupils filter out extraneous material before passing this on to the working space of the mind. When too much information passes into the working space, it is unable to process it and it 'crashes'. Chemistry lessons frequently introduce abstract concepts alongside complicated laboratory procedures, so it is easy to overload the working space. Johnstone recommends that the teacher of chemistry should consider how best to present this information in more

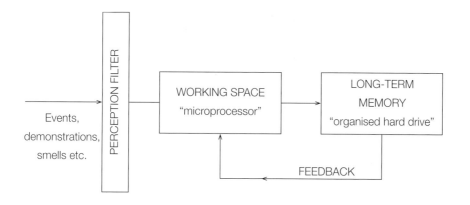

manageable chunks through clear explanations and useful analogies, to ensure that when information is processed successfully, it is passed to the long-term storage area for retrieval at a later time. Teachers of chemistry could try to point out patterns in chemical ideas so that the pupils' information 'filing system' in the long-term memory can be accessed and used easily.

The role of the teacher of chemistry is also to enable all the pupils to learn chemistry in a safe and supportive environment. Safety is paramount and the following advice from CLEAPPS, the School Science Service at Brunel University, should **always** be adhered to.

The chemistry laboratory is a potentially dangerous place and before planning any activity you should:

1 Make sure you are aware of the chemistry of any reactions taking place.

2 Be aware of possible hazards.

3 Practice any difficult procedures yourself beforehand.

4 Carry out a risk assessment.

5 Stick to your tried-and-tested plan in the lesson.

Chemistry has the advantage of hands-on practical lessons and awe-inspiring concepts. However the enthusiastic teacher of chemistry should be aware that some pupils in their classes may not be quite so awe-struck about molecules and that it will require expert pedagogical skills and practice to maintain their interest.

In summary, Alex Johnstone offers the following 'Ten Education Commandments' to teachers of chemistry:

1 What you learn is controlled by what you already know.

2 How you learn is controlled by how you have learned successfully in the past.

3 If learning is to be managed, it has to link to existing knowledge and skills, enriching and extending both.

4 The amount of material to be processed in unit time is limited.

5 Feedback and reassurance are necessary for comfortable learning, and assessment should be humane.

6 Cognisance should be taken of learning styles and motivation.

7 Students should consolidate their learning by asking themselves about what is going on in their own heads.

8 There should be room for problem solving in its fullest sense to exercise and strengthen linkages.

9 There should be room to create, defend, try-out and make predictions.

10 There should be opportunity given to teach (you don't really learn till you teach).

These commandments have been influential in my own teaching and in drawing up and using the activities and suggestions set out in the rest of the book.

References

Johnstone, A.H. (1997) Chemistry teaching, science or alchemy? *Journal of Chemical Education* **74** (3): 262–281

Shayer, M. and Adey, P. (1981) *Toward a science of science teaching.* Heinemann, London

1 Classifying Materials

chapter 1

Solids, liquids, gases and the particle theory

BACKGROUND

The particle model of solids, liquids and gases is often introduced at KS2 with further development at KS3 and foundation level GCSE (see Table 1.1).

It is important to reinforce to the pupils that the particle theory represents a microscopic model which helps to explain macroscopic properties of materials. There is a danger that these models are presented to pupils as how things really are, so that pupils come to believe that all materials really look like the 2D static

Table 1.1 Progression through the Key Stages

KS3 Level 3–5	KS3 Levels 5–8/ Foundation GCSE	Higher GCSE

• Recognise the difference between solids, liquids and gases
  – density
  – compressibility
  – ease of flow
  – shape/volume
• Particle theory
• Melting/dissolving
	• Arrangement of particles into solids, liquids and gases
	• Kinetic theory, changing state
		• Structure and arrangement of atoms
		• Structure and how this relates to the physical properties
		• Solubility curves

diagrams used in textbooks to illustrate the concept. The particle model first introduced at Key Stage 2 will be refined and adapted as the pupil progresses through the key stages from a 'billiard ball' particle idea towards a simplified solar system model of the atom through valence shell repulsion theory and sub-atomic particles at A-level. It is unhelpful to present any of the models as 'how it actually is' only to discredit this at the next stage of learning. It is better to say that the model will help answer questions at this particular stage of learning.

The sheer size and numbers introduced can often be rather daunting to younger pupils and it might help to use a macroscopic frame of reference to start with. Water drops are familiar and the story which follows (with apologies to Richard Feynman) could be used to set the scene: A water drop is probably less than 1 cm long. When this is magnified about 2000 times, the drop will then be roughly 20 m; it still looks smooth. If you look carefully, however, you might see football-shaped objects swimming around. This is the domain of the biologist who will want to stop the magnification and study the behaviour of the wiggling bodies. The physical scientist will want to carry on. If the water is then magnified a further 2000 times, the drop will now extend for 24 km and on close inspection you will see a kind of teeming movement. The water is no longer smooth in appearance and looks like a football game from a great distance. On further magnification, individual blobs can be seen. The diagrams which are used in textbooks to illustrate these blobs are idealised.

- The particles are drawn in a simple manner with sharp edges which is inaccurate.
- They are in two dimensions, but of course the particles are moving around in three dimensions.
- Symbols are used to represent different kinds of blobs.
- The diagrams are static when the particles are in fact in constant motion. Gas particles move very quickly between collisions at an average velocity of 500 m/s.
- The diagrams are unable to represent that the blobs are stuck together.
- Many suggest disproportionately large gaps between particles.

If an apple was enlarged to the size of the Earth, the particles or atoms in the apple would be about the size of the original apple compared to the Earth. There are probably around 5×10^{20} particles in a matchbox full of air.

Let's go back to the water droplet with its jiggling particles stuck together and tagging along with each other. The water retains its volume and will flow because of the attraction of the molecules for each other. (The concept of a molecule is approximate and exists only for a certain class of substances. A water molecule contains two hydrogen atoms stuck to one oxygen atom, however crystalline solids such as salt are not referred to as molecules of salt.) Heat energy is

Table 1.2 Particle model acceptable at KS3 and Foundation Level GCSE

	Solids	Liquids	Gases
Kinetic energy of particles	Low	Low	High
Motion of particles	Vibrate about fixed positions	Freely around one another	Ceaseless and random
Rate of diffusion	Very slow	Slow	Fast
Attractive forces	Strong	Fairly weak	Negligible
Spacing of particles	Close contact	Close contact	'Far' apart
Arrangement of particles	Regular	Random	Random
Volume	Fixed	Fixed	Variable
Shape	Fixed	Variable	Variable
Compressibility	Virtually incompressible	Only slightly compressible	Very compressible
Density	High	High	Very low

represented as changes in the speed of the jiggling motion. As the temperature increases, the motion increases and the volume occupied increases as the distance between the particles increases. When the pull between the molecules is insufficient to hold them together, the particles fly apart and become separated from each other – the water changes to steam. Under the right conditions particles of air jiggling around quickly above water can bump into water molecules and knock them into the air. It will then be juggled around by air particles i.e. the water evaporates. Diagrams of gases often illustrate many gas particles in a small volume when in reality there may only be a few molecules in a whole room. Gas molecules bounce against the walls, exercising jittery pushing forces. If gases are squashed into smaller spaces, this average push or pressure increases.

If the jiggling motion is decreased in liquids, by reducing the temperature, the particles will lock into a new pattern with each particle arranged in an array or crystalline lattice – i.e. a solid.

When crystalline solids of salt are added to water they dissolve. The crystal lattice of salt is made up of particles which are not strictly speaking atoms. They are in fact charged atoms or ions. The ions are stuck together in the lattice because of the electrical attraction between them. When the solid is added to water, the attraction of the negative charge on the oxygen part of the water molecule and the small positive charge on the hydrogen part of the water jiggles the ions loose from the lattice.

KEY STAGE 3 CONCEPTS

Extensive research has been carried out into children's alternative concepts and in this area by the Childrens' Learning in Science project, CLIS, based at Leeds University. Their research has shown that children have already developed ideas about particles and that some of them hold alternative interpretations. For example, they:

- do not make the link between the microscopic model and the macroscopic properties observed.
- do not appreciate that it is the bombardment of particles on a vessel wall which accounts for the pressure and temperature inside the vessel.
- attribute macroscopic properties to the particles, for example suggesting it is the swelling of the particles which causes expansion or that the particles themselves melt.
- suggest that it is the changes in force between the particles which alter temperature and pressure.
- think that changing state is a continuous process.
- hold the view that particles stop moving on cooling and becoming a solid.
- do not grasp the idea that there is a vacuum in the gaps between particles or that particles move about more and occupy a large volume when heated.

All pupils arriving in your KS3 classes will have already studied solids, liquids and gases at KS2. Unfortunately, many of the long-standing commercial schemes of work do not acknowledge this, and often lessons in year seven simply repeat work already done. This is very demotivating for some pupils and so it is crucial that you establish what pupils already know about this area.

There are a number of ways you could do this, for example, by asking pupils to

- draw a concept map
- carry out a card sort of key definitions
- answer probing questions such as those found in the CLIS materials.

Having established what pupils already know, it is then possible to set up a number of activities which challenge alternative conceptions and help develop a more 'accepted' mental model.

Activities 1 to 4 in Figure 1.2 are intended to help pupils gather evidence that suggests that particles in solids are arranged in regular patterns and are constantly moving. Activities 5 to 7 look for evidence about the properties of liquids. Activities 9 and 10 look at properties of gases.

> ## Assessing pupils' learning
>
> Ask pupils to:
>
> - draw a series of pictures, one to show how they imagine the particles are arranged in a solid, another for a liquid, and a third for a gas.
> - describe what happens to the particles in a block of ice when it is taken out of a freezer at $-10°C$ and placed in a fridge at $-1°C$.

Dissolving and melting

A model of dissolving which is appropriate at this stage is that

- not all solids dissolve. Some are said to be insoluble.
- the solid does not disappear but dissolves and is still in the solution.

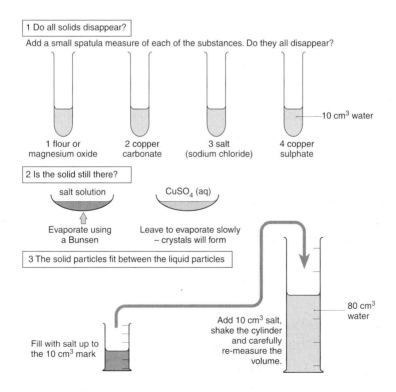

Figure 1.1 *Pupils' ideas about dissolving*

1 Packing of particles / density of solids

materials kit

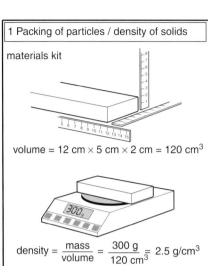

volume = 12 cm × 5 cm × 2 cm = 120 cm^3

density = $\dfrac{\text{mass}}{\text{volume}}$ = $\dfrac{300\ \text{g}}{120\ \text{cm}^3}$ = 2.5 g/cm^3

2 Liquids solidifying into regular structures

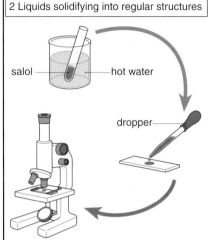

salol — hot water

dropper —

3 Regular solid structures / crystals

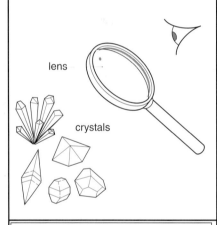

lens

crystals

4 Diffusion in solids

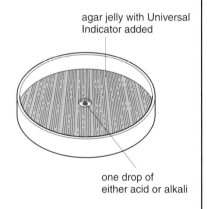

agar jelly with Universal Indicator added

one drop of either acid or alkali

5 Particles fit into spaces

1 Fill each beaker to overflowing with water
2 Then drop water or ethanol carefully

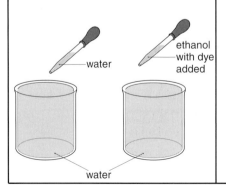

water

ethanol with dye added

water

6 Using a model

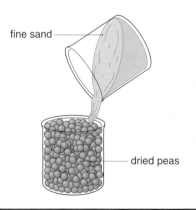

fine sand

dried peas

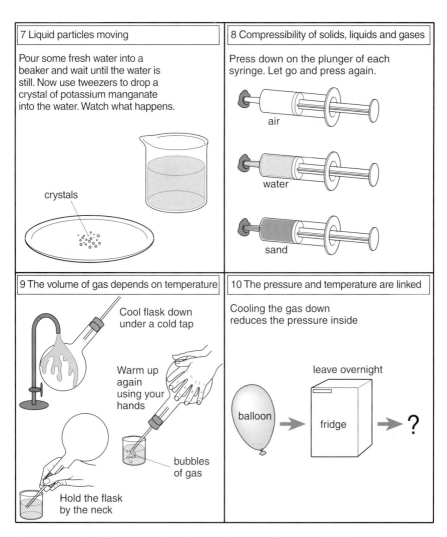

7 Liquid particles moving	8 Compressibility of solids, liquids and gases
Pour some fresh water into a beaker and wait until the water is still. Now use tweezers to drop a crystal of potassium manganate into the water. Watch what happens. crystals	Press down on the plunger of each syringe. Let go and press again. air water sand
9 The volume of gas depends on temperature	10 The pressure and temperature are linked
Cool flask down under a cold tap Warm up again using your hands bubbles of gas Hold the flask by the neck	Cooling the gas down reduces the pressure inside leave overnight balloon → fridge → ?

Figure 1.2 *Particles – suggestions for activities to challenge pupils' ideas*

- the solid (solute) is broken down into smaller particles which fit between the liquid (solvent) particles.
- the solute and solvent together make the solution.

The activities in Figure 1.1 could be used to reinforce these ideas. Activity 1 helps clarify the difference between disappearing and dissolving as the blue colour of the copper sulfate remains. Pupils will know that adding sugar to a liquid makes it sweet to taste and that the sugar can be recovered if the tea is evaporated.

Assessing pupils' learning

Ask pupils to:

- draw a particle cartoon sequence to show what happens when a sugar cube is dropped into a cup of tea.
- write an explanation of the difference between melting and dissolving for a Y5 or Y6 pupil.
- label a diagram like Figure 1.3 of a bottle of lemon squash to help develop their language skills.

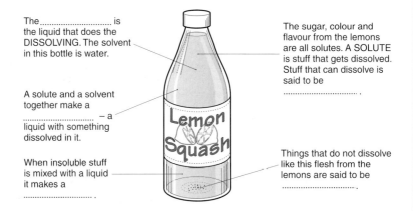

The.............................. is the liquid that does the DISSOLVING. The solvent in this bottle is water.

A solute and a solvent together make a .... – a liquid with something dissolved in it.

When insoluble stuff is mixed with a liquid it makes a ....

The sugar, colour and flavour from the lemons are all solutes. A SOLUTE is stuff that gets dissolved. Stuff that can dissolve is said to be ....

Things that do not dissolve like this flesh from the lemons are said to be ....

Figure 1.3 *Assessing pupils' learning – dissolving*

KEY STAGE 4 CONCEPTS

Colloids

A substance is defined as a **colloid** if the particles dispersed in the continuous phase are between 1 mm and 100 mm across. They are intermediate between coarse suspensions like milk of magnesia and solutions like copper sulfate or vinegar.

Colloids can be identified by shining a beam of light through them. Colloidal particles reflect light back. You could demonstrate this by shining a beam of light through vinegar and orange squash.

Every colloid has two parts: for example, the continuous phase is the water part of orange squash, the disperse phase is the orange particles scattered throughout the water. In this context the word phase is used to describe two substances present but separated by a clear boundary. Table 1.3 shows the different types of colloids.

Table 1.3 Colloids

Continuous phase	Disperse phase	Type	Example
Gas	Gas	–	–
Gas	Liquid	Aerosol	Mist
Gas	Solid	Aerosol	Smoke
Liquid	Gas	Foam	Whipped cream
Liquid	Liquid	Emulsion	Hand cream
Liquid	Solid	Sol	Paint
Solid	Gas	Solid foam	Pumice
Solid	Liquid	Solid emulsion	Butter
Solid	Solid	Solid sol	Pearl

Linking the microscopic and the macroscopic

Atomic structure is introduced at KS4 foundation level (see 'Atomic Structure'), however only higher candidates need to be able to link the arrangement of the particles with the physical properties of the material.

You could challenge the pupils by asking them to find out the following.

- Why does salt dissolve, and why does it appear to be cubic shaped?
- Why does iodine sublime rather than melt and boil?
- Why do metals form crystals and conduct electricity?
- What happens when metals are beaten into shapes and pulled out into wire?
- Why, if they are both made of carbon, is diamond so hard and graphite so soft?
- Small molecules are gases at room temperature, why then is water a liquid at room temperature?

The physical properties such as melting point, boiling point, ability to conduct and appearance are linked to the arrangement of atoms. Pupils could test the conductivity of copper, sodium chloride solid and sodium chloride solution.

Safety Advice: You should demonstrate the heating of iodine. See CLEAPPS Hazcards for details.

As a general rule, small molecules are gases at room temperature and do not conduct; larger structures are solids and some conduct when free electrons are available.

Water should be a gas at room temperature but is a liquid because of the hydrogen bonds between the water molecules. Iodine sublimes because the weak bonds holding the diatomic molecules together in the regular arrangement of the solids break on gentle heating producing a purple vapour which is made up of small diatomic molecules of iodine. Sodium chloride and copper are large structures arranged in lattices. There are no free electrons available in sodium chloride because they are bound up in the orbits of the atoms in the lattice. They can be released as ions in the molten state or in solution, therefore, sodium chloride conducts electricity. Copper metal has a face-centred, cubic close-packed structure. Bonding in metals is generally considered to be by the loss of one or more electrons which form a molecular orbital surrounding the close-packed metal ions. This is often referred to as 'a sea of electrons'. The good conductivity of metals can be explained by assuming that this 'sea of electrons' allows current to flow easily (see 'Useful Products from Metals').

Graphite and diamond are both allotropes of carbon. Buckminster fullerene is a recently discovered third allotrope. Allotropes are the same element but with a different structural arrangement of the atoms. The *Chemistry Set* CD-ROM from New Media includes excellent molecular models of carbon and most textbooks will include good diagrams. Diamond is the strongest and most unreactive form of carbon in which atoms are joined together in a repeated tetrahedral arrangement in a giant structure. There are no weak areas because all the bonds are equally strong. Graphite, on the other hand, is arranged in sheets of carbon atoms. Each carbon atom in the sheet is attached to three other carbons. This means that one carbon has free electrons not involved in the bonding process and so graphite is the only non-metal which can conduct. There are weak forces between the sheets which enable them to slide over each other. This can be demonstrated by using a 9B pencil which is almost pure graphite and very soft.

Going further

More able pupils could research the three allotropes of carbon – graphite, diamond and Buckminster fullerene – using model kits or the *Chemistry Set* CD-ROM and present a Powerpoint slide show to the class.

Table 1.4 Summary of structures. (Oxygen and iodine are represented by ball-and-stick models, while water is a space-filling model)

Material	Melting/boiling points	State at room temperature	Bond strength	Structure	Diagram
Oxygen	Low	Gas	Strong bonds between atoms; weak bonds between molecules	Small molecules	
Water	Low	Liquid (should be a gas)	Strong bonds between atoms; hydrogen bonding between molecules	Small molecules disordered; close together and free to move about	
Iodine	Low	Solid	Strong bonds between atoms; weak bonds between molecules. Sublimes at low temperatures	Small molecules in vapour; low melting point ordered solid	I_2 vapour / I_2 solid / weak attraction
Sodium chloride	High	Solid	Strong bonds	Giant structure of ions in a regular lattice	Na Na Na
Copper metal	High	Solid	Strong metallic bonds	Giant regular close packed solid	

Assessing pupils' learning

Confirm that pupils have grasped the key ideas by asking them to match the properties and structures of the materials in Figure 1.4.

1 Descriptions:

small molecules ordered together	single atoms disordered, fast moving and far apart	close-packed giant structure of metal atoms	giant structure of ions	giant structure of non-metal atoms
disordered giant structure	small molecules disordered, close together and free to move about			

2 Properties

a shiny, bendable solid with a high melting point, conducts electricity when solid and when liquid	a gas at room temperature and pressure	a liquid at room temperature and pressure	very hard, very high melting point does not conduct as a solid or gas	high melting point solid, conducts electricity when liquid but not when solid
a non-crystalline supercooled liquid which is solid at room temperature and does not conduct when solid	a low melting point solid, sublimes on gentle heating, does not conduct electricity when solid or liquid			

3 Examples

glass	helium	copper metal	magnesium chloride	water	diamond	iodine

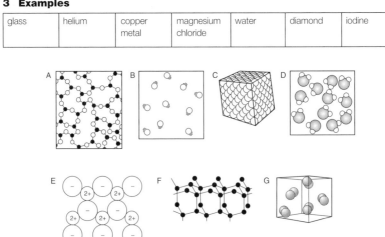

Figure 1.4 *Properties, structure and bonding*

New materials – ceramics

Ceramics are a class of materials made from inorganic compounds by high temperature processing. They are resistant to chemical attack and include glasses, cement, pottery, china and porcelain.

Glass is a supercooled liquid which forms a non-crystalline solid. Common lime-soda glass is a mixture of sand, sodium carbonate and calcium carbonate. Glass is usually coloured by adding metal oxides, and the deep red colour in old stained glass is achieved by using gold.

The Nuffield *Co-ordinated Chemistry* Teacher's Guide has a good section on making and blowing glass.

New materials – composites

Many of the high performance materials used particularly in sports equipment are made from composite structures in which two or more materials are combined together. This idea was first used by the early metallurgists who combined metals with other metals or charcoal from their fires to make alloys. Many of the new materials mimic naturally-occurring biological systems like bone and wood. By combining more than one material, the range of materials available to the designer can be increased. Many composites are high-performance materials consisting of a plastic matrix reinforced by fibres like glass, graphite or boron. In addition to sports goods, they are widely used in space travel, aircraft and car components. Figure 1.5 shows how engineers can make composites.

> ### Assessing pupils' learning
>
> Ask pupils to carry out a survey of the composite materials used in sporting equipment and to investigate and explain how the structure improves performance.

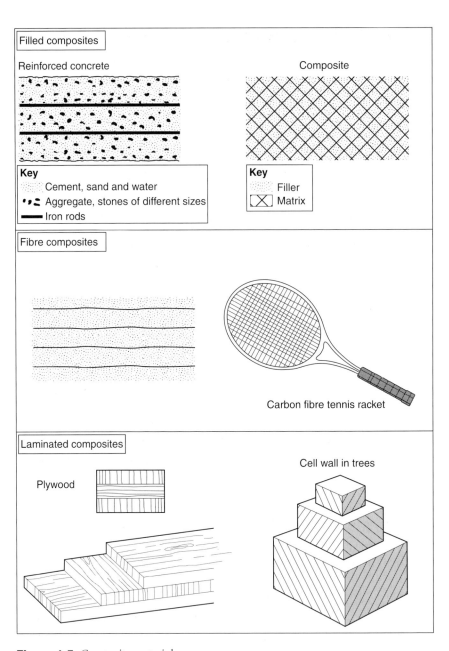

Figure 1.5 *Composite materials*

ENRICHMENT AND EXTENSION ACTIVITIES

Information communication technology

KS3

Netherhall Software *Moving Molecules* (BBC) models the motion of particles and simulates melting and boiling. It can be used to investigate the relationship between temperature and pressure.

The States of Matter CD-ROM from New Media has written support material and provides really good simulations of melting, boiling and evaporating.

Datalogging – plot cooling curves of octadenoic acid to monitor changes of state.

KS4

Use a word processing package and scanner to illustrate a presentation on the properties and structures of materials. The Chemistry Set CD-ROM from New Media includes a large database of molecular structures.

Search the Internet for up-to-date information on new materials such as Gortex™, Lycra™, Kevlar™ and carbon re-enforced fibres, and produce an information pamphlet.

Use electronic mail and resources from the Science Across the World project (http://www.bp.com/saw/) such as 'Drinking Water'. Enrol as a partner school and e-mail schools throughout the world with your information about local drinking water.

Use Excel to plot solubility curves.

Practical activities and investigations

During the Second World War the Allied Forces tried to use icebergs strengthened with straw as large aircraft carriers in the North Atlantic. Plan an investigation to simulate this on a smaller scale using ice cubes. Which is the best readily-available material which strengthens the ice? (Cotton wool, iron wool or straw work quite well.)

Survey and test the properties of materials used to fill teeth. See the GNVQ module Dental Dilemma, from the CIEC centre at York, for details.

Make a stabilised emulsion in the lab using egg whites etc. (See any cookery book for the recipe for mayonnaise.)

Investigate the viscosity of liquids over a range of temperatures.

Make slime – see *Salters' Chemistry Club Handbook*.

References

Video

The Chemistry of Almost Everything – Survival 4 – materials used in the space shuttle. ST240, The Open University

Science of New Materials, BBC Video

Science in Action: Mixtures, BBC Video

The Works: Fluids, BBC education and training video. This excellent video considers the anomalous properties of water, colloids and oil.

Teacher resources

Investigations with Water – coloured pupils' booklets with practical activities and background information with a helpful teacher's guide. Available free from BNFL.

Fit to Drink – a photocopiable resource pack with water analysis instructions and ion-exchange resin available from CIEC at York.

GNVQ modules on materials – *Dental Dilemmas* available from CIEC York.

A Question of Cooling – using the techniques of refrigeration and commercial freezing to help pupils understand changing state (11–14), CIEC York.

Sweet Success – a study of solutions and crystallisation (13–14), CIEC York.

Clean Science – a theory of detergency and stain removal (11–14), CIEC York.

What's the Solution – introduces ideas about solutions and solubility (13–14), CIEC York.

Books

Atkins, P. *Molecules*. Scientific American Library

Hill, G. *Making New Materials*. Hodder & Stoughton, London

Gordon, J. *New Science of Strong Materials*. Penguin, London

Feynman, R. *Six Easy Pieces*. Penguin, London

chapter 2

Elements, compounds and mixtures

BACKGROUND

One of the greatest achievements of chemistry has been to show that all matter in the universe, be it a super nova, a starship enterprise, planet Earth or a year seven pupil, is built up from a combination of about 100 elements.

Elements cannot be broken down into simpler substances by heating, boiling, adding acid or any other method available to the chemist to change a substance. Atkins suggests that to go further requires the aggressive techniques of the physicists who 'can smash elements apart into electrons, protons and the other fundamental particles using high energy particle accelerators'. The smallest amount of an element that can exist is an **atom**. An element, such as metallic gold, is a collection of identical atoms and is identified by its **atomic number**, which is the number of protons in the nucleus (see 'Atomic Structure'). For example, only a carbon atom has six protons in its nucleus.

A **compound** is a combination of elements – water is a combination of hydrogen and oxygen, and aspirin is made up of carbon, hydrogen and oxygen. Many compounds consist of molecules which are discrete groupings of atoms in a definite geometrical arrangement. A suitable explanation at Key Stage 3 is that a compound contains two or more elements bonded together in fixed proportions. A compound usually has different properties from the elements from which it is made. For example sodium metal is a reactive metal and chlorine a toxic gas, but combined together they make sodium chloride which is added to food to preserve it or to enhance its flavour. Substances such as glass, steel, iron oxide, plastics and starch are compounds too but the proportions of elements making them up can vary. In each case atoms are attracted by all the adjacent atoms making them nearly uniform solids. **Alloys** are also compounds.

Mixtures contain elements or compounds in varying proportions. The components of mixtures have different physical properties and so can be separated. Ink, for example, can be separated into its components because the solvent used – water – boils at relatively low temperatures while the pigments added to colour the ink have very high melting and boiling points. It is possible to boil off the solvent at 100°C leaving the dyes in the ink behind. Nearly everything we eat is a mixture. For example, orange juice contains the compounds water, sugar and citric acid. Other mixtures include: air which is a mixture of gases and water vapour; sea water which is a mixture of water, sodium chloride and magnesium chloride; and crude oil which is a mixture of hydrocarbons with different boiling points (see 'Crude Oil').

KEY STAGE 3 CONCEPTS

Pupils are often able to quote a textbook definition of the word 'element' and not able to use a mental model of the concept. Frequently too, pupils assign macroscopic properties to individual atoms and are, for example, unable to use the idea that copper atoms have very different properties from those of the copper element.

It is a good idea to set the subject in context and a way of doing this would be to ask pupils to research details about one specific element leading to the production of a class periodic table. Pupils could find out the following information about their chosen element:

- date of discovery
- where it is found
- how abundant it is
- atomic number
- appearance
- what it is used for
- other interesting or unusual information.

Information could also be collected from a CD-ROM or periodic table web pages. Pupils could produce a pop-up card about the element or give an oral presentation to the group.

Able pupils could be asked to write about the life of the carbon atoms they have just breathed out or to write a poem about their chosen element.

It is important that adequate time is spent introducing these key concepts and that they are reinforced before pupils encounter chemical reactions. A demonstration of the formation of iron sulfide provides a good introduction. It is good practice to think carefully about which is the most appropriate teaching style to ensure that learning takes place. The main objective at this stage will probably be to establish the concept by helping pupils to develop mental models

Table 2.1 Teacher demonstration of the differences in properties between iron and sulfur and iron sulfide

Name	Appearance	Is it magnetic?	Does it react with acid?	What happens when it is heated?
Iron (Fe)	Grey/black granules	Yes	Slight bubbling	No reaction
Sulfur (S)	Yellow powder	No	No reaction	Melts into an amber-coloured viscous liquid
Iron/sulfur mixture	Speckled yellow/ grey powder	Iron separates from the mixture	Slight bubbling around the iron, sulfur does not react	Glows red
Iron sulfide	Black solid	No	Vigorous bubbling; hydrogen sulfide gas given off which smells like rotten eggs	No reaction

of elements and compounds. Practical skills can be better developed in other topics so in this case a teacher demonstration might be more effective.

It is good practice to try to involve the pupils in the demonstration. You could do this by asking them to complete a table as they watch and by asking pupils to assist you with parts of the demonstration.

S Safety Advice: You will need to carry out a risk assessment before you start.

Start by heating the iron/sulfur mixture in a test tube over a hot flame until it glows. Then remove the tube from the heat source because the reaction is exothermic. Allow the test tube to cool down while you give an explanation of what elements and mixtures are using models – you could use model kits or coloured LEGO™ blocks. When the test tube has cooled down, wrap it in scrap paper and break the tube with a hammer. Extract the solid lump of iron sulfide carefully and ask a pupil to describe its appearance, carry out the same tests on the lump and complete the table. It is helpful then if you outline the rules for naming compounds:

1 The name of the metal comes first and is followed by the name of the non-metal. The ending of the non-metal changes depending on which elements

are present in the compound.

$$\text{iron} + \text{sulfur} \rightarrow \text{iron sulfide}$$

2 Compound ending

-ide Only two elements are present in the compound e.g. iron ox*ide* is made up of iron and oxygen.

-ate Three elements are present in the compound including oxygen e.g. copper sulf*ate* contains copper, sulfur and oxygen.

-ite As above, there are three elements including oxygen, although -ite compounds contain less oxygen than -ate compounds. For example, sulfate, SO_4, and sulfite, SO_3, contain the same elements but the latter has less oxygen.

3 Hydrogen compounds

If a compound contains hydrogen, the non-metal part of the compound name may start with the prefix hydr-.

Assessing pupils' learning

Ask pupils to:

- write a definition of an element, a compound and a mixture in their own words aimed at younger pupils.
- complete Table 2.2. The first row has been completed to start them off.

Table 2.2

Name	Elements
Iron oxide	Iron and oxygen
Sodium sulfate	
Sodium oxide	
Potassium hydroxide	
Calcium carbonate	

- complete the gaps in the following passage using the periodic table.

'Elementary Inspector'

The burglar had been caught red-handed with the (Ag) _____ and (Au) _____. He had been caught after a high speed (carbon + argon) _____ chase. The inspector was anxious to get this one (Fe) _____ed out before the end of the night. The victim arrived at the station shouting (holmium tungsten) _____ dare someone (sulfur + tellurium + aluminium) _____ from me. The (cobalt) _____nstable (potassium + neon + tungsten) _____ this was not going to be an easy (calcium + selenium) _____ to solve.

Separating mixtures

Pupils will have separated mixtures at KS2 but will probably have used everyday utensils like colanders. It is important that pupils are shown how to carry out the separation techniques correctly. You could demonstrate how to carry out each technique, explaining clearly how the separation is achieved. Figure 2.1 summarises some separation techniques commonly used at KS3.

Research carried out about how children learn suggests that it is best to introduce concepts, skills and understanding in small stages and to develop problem-solving activities which bring together all the strands as a method of establishing deeper understanding (see Figure 2.2).

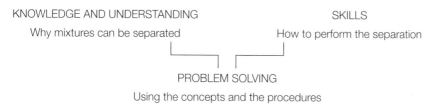

KNOWLEDGE AND UNDERSTANDING | SKILLS

Why mixtures can be separated | How to perform the separation

PROBLEM SOLVING

Using the concepts and the procedures

Figure 2.2 *Establishing deeper understanding*

When you have introduced each technique, issue each pupil with a separating mixture problem (see Table 2.3). Ask them to carry out the separation and to write a brief report outlining what they have done. More able pupils could be given the more difficult problems.

Assessing pupils' learning

One method of checking that pupils have understood the technology of separating mixtures and have a written record of this would be to ask them to produce their own technical guide to separating mixtures. The audience for their guide could be Y6 pupils and it should show the techniques available for separating mixtures and why they work.

Mixtures can be separated because the constituents have different physical properties

Decanting is the quickest way of separating an insoluble substance from a liquid. Mud settles to the bottom of a river because it is denser than water. Decanters were used to store port which often contains insoluble particles.

Mud is insoluble in water

Filtering is a much more precise way of separating an insoluble substance from a liquid. Filter paper has lots of tiny holes in it which are big enough to allow water through but stop bigger particles. Coffee filters prevent large coffee granules from getting into the cup.

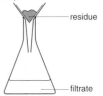

residue

filtrate

Ground coffee is insoluble in water.

When solutions containing soluble solutes are **evaporated**, the liquid evaporates leaving the solid behind. Salt can be separated in this way, and in Israel large amounts of sea water are left to evaporate in huge ponds.

solution

Salt is soluble and dissolves in water. It cannot be filtered because the salt particles are smaller than the holes in the filter paper. Water boils at a much lower temperature than salt.

Gin and tonic mix completely. They are said to be **miscible**. Gin is an alcohol molecule. **Distillation** is used to separate the two liquids. Fractional distillation is used to separate the many miscible liquids found in crude oil.

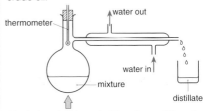

thermometer

water out

water in

mixture

distillate

Alcohol and water mix. They have different boiling points so it is possible to separate them by heating up the mixture until the alcohol boils and the vapour can be condensed as it leaves the water behind.

Some liquids do not mix very well. Vinegar and oil do not mix and separate into layers if they are left to stand. A special **separating funnel** with a tap at the bottom is used to separate immiscible liquids.

oil

vinegar

Oil and water have different physical properties; they do not mix well together. Oil is less dense than water and will float on top of it.

Paper chromatography is used to separate dyes. Ink pigments are made up of a range of different dyes. The dyes separate because some of them are carried through the paper quicker by the solvent.

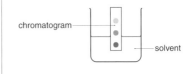

chromatogram

solvent

Figure 2.1 *Separating mixtures*

Table 2.3 Separating mixtures problems

1   Oops! I have accidentally poured oil into my ink bottle. Can you help?

2   There have been five murders in a small Cambridge hotel. Each murder victim has been stabbed and poems written in black ink have been found attached to the bodies. The murderer was disturbed and dropped his pen. The hotel has been searched and two bottles of black ink have been found. They belong to Professor Kram A.T. O'Gram and Dame Eva Poration. Can you solve the mystery?

3   I have picked up the wrong coffee jar. I meant to add instant coffee to the cup but have used Dad's really expensive ground coffee instead. Help!

4   Smarties™ manufacturers suspect that cheap imitations are being passed off as their brand by unscrupulous suppliers. Can you investigate?

5   Dr Liebig has just discovered a rock salt deposit. It is very impure. She would like to prepare pure white samples of salt from the sample taken from the deposit.

6   Mrs Spinks has a problem – her class. They are very messy. Last week she had three jars, one with sawdust, one with sand and a third one with salt. Now they are all mixed up! Please help.

ENRICHMENT AND EXTENSION ACTIVITIES

Information communication technology

CD-ROM
Elements, Compounds and Mixtures from New Media
Chemistry Set – Periodic Table Database from New Media
Materials from YITM

Desk top publishing
Separating mixtures technical guide using Word, Publisher or similar package. Import scanned images from textbooks.

Internet
Search web periodic tables:
http://www.shef.ac.uk/~chem/web-elements/html
http://www.chemsoc.org/viselements

Practical activities and investigations

Separating mixtures problems and chemical egg races in *In Search of Solutions* published by the Royal Society of Chemistry.

References

Videos

Atoms, educational activities videos available from Boulton–Hawker Films Ltd

Discovering Elements, S102, The Open University

The Strange Story of Napoleon's Wallpaper, Human Element from BBC Video for education and training

ST240 – The Chemistry of Everything Almost, The Open University

Elements, Compounds and Mixtures, Viewtech Film and Video

Science in Action: the Periodic Table, BBC Video

Teacher resources

SATIS 1209 *Are you made of stardust?* This unit considers the formation of elements following the explosion of supernovae.

Books

Atkins, P. (1984) *Molecules.* Scientific American Library

Emsley, J. (1987) *The Elements.* Oxford University Press, Oxford

Emsley, J. (1987) *An A to Z of the Elements.* Channel 4 Television to accompany *Equinox: The Elements* video

Cox, P. *The Elements; their Origin, Abundance and Distribution*

chapter 3

Atomic structure and bonding

BACKGROUND

A good understanding of atoms and how they bond together is important for the prospective chemist. For the unenthusiastic too, if presented sensitively, the story of how they were discovered might just be retained. Feynman summed up how important this concept is when he said

'If, in some cataclysm, all of scientific knowledge were to be destroyed, and only one sentence passed on to the next generation of creatures, what statement would contain the most information in the fewest words? I believe it is the atomic hypothesis (or the atomic fact, or whatever you wish to call it) that all things are made of atoms, little particles that move around in perpetual motion, attracting each other when they are little distance apart, but repelling upon being squeezed into one another. In that one sentence, you will see, there is an enormous amount of information about the world, if just a little imagination and thinking are applied.'

The word atom comes from the Greek word *atomos* meaning indivisible particle, and the Greek Philosophers are credited with the first atomic theory. We now know that atoms can in fact be broken down into sub-atomic particles. A more precise definition might be that elements can be sub-divided into atoms and that although further sub-division can take place, the identity of the original element will be lost.

At KS3 and KS4 pupils are introduced to models of atomic structure which have been developed over a number of centuries. The Rutherford 'solar system' model introduced at Key Stage 4 will enable pupils to explain why chemical reactions take place. The model will be refined at A-level and beyond, when more complex ideas will be introduced. Atomic structure is included in both single and double award syllabuses.

KEY STAGE 3 CONCEPTS

Some pupils find this area quite difficult. This may be because they are asked to use models of sub-atomic systems to account for macroscopic events. The representation of atoms in the form of dot and cross diagrams further compounds this. Textbooks very rarely introduce the diagrams as one model of how to interpret the structure of the atom, and as a consequence many pupils hold the view that atoms actually look like the textbook diagrams. If probed, some pupils describe the physical properties of an atom as the same as that of the bulk material – so copper atoms are described as good conductors with a high melting point.

To help pupils grasp the concepts it will help if they have already met the idea of an element before being introduced to atoms, and they are encouraged to think of atoms in three dimensions. The CD-ROM *Atom Viewer* shows atomic structure by using shell diagrams, dot and cross diagrams and three-dimensional orbiting electrons. It also helps make the subject relevant if it can be seen in an historical context and be given a human face. The remarkable story of the discovery of the atom is lost when it is presented in textbooks as a fait accompli. Pupils who are not enthused by abstruse concepts might respond to the story of the discovery of the atom being told (Figure 3.1). There are a number of ways this could be done:

- individual pupils might be given a period of history or a scientist to research and be asked to produce a small display for a class diorama.
- each pupil could be issued with a small card containing information about one scientist who has contributed to the discovery of the atom. The class could then arrange themselves in chronological order and then explain the contribution of the scientist or event. This could be the basis of a timeline display. Figure 3.1 provides background information.

Able pupils might be asked to research the history of the atom from Democritus to the present day and present this as a Powerpoint presentation, or other media form.

KEY STAGE 4 CONCEPTS

You could establish what ideas pupils already have about atoms by using concept maps or by asking the following questions:

1 What do you think you would see if you magnified a pile of sulfur powder as much as possible?

2 Describe the properties of individual iron atoms.

3 Imagine you have a pair of Superman's spectacles which enable you to see inside copper foil. Draw what you think you would see.

420BC	Ancient Greece	1919	Aston (Cambridge)
	• indivisible particles – atoms		• invents mass spectrometer and discovers non-radioactive isotopes
1660s	Newton (Cambridge)		
	• supports atomists		
1808	Dalton (Manchester)	1919	Rutherford
	• founder of atomic theory		• artificial disintegration of the atom
1895	Roentgen (Germany)		
	• discovers X-rays	1920	Bohr (Copenhagen, Denmark)
1896	Becquerel (France)		• opens Institute for Theoretical Physics
	• discovers radioactivity		
1897	JJ Thomson (Cambridge)	1920	Rutherford
	• discovers unit carrier of negative electricity (electron)		• identifies proton (hydrogen nucleus) as basic constituent of all atomic nuclei
1898	Curie (France)		
	• discovers radiation in pitchblende	1923	Compton (Chicago)
			• Compton effect
	• extracts radium	1925	de Broglie (France)
1900	Thomson/Rutherford (Cambridge)		• wave theory of light
	• show that Becquerel radiation is made up of α, β and γ radiation	1926	Schrödinger (Germany)
			• wave equation
1901	Planck (Germany)	1927	Heisenberg (Copenhagen, Denmark)
	• quantum theory (discontinuity of energy)		• uncertainty principle
		1927	GP Thomson
1902	Rutherford/Soddy (Canada)		• wave-particle duality
	• spontaneous disintegration of some atoms	1928	Dirac (Cambridge)
			• β-particle problem
1905	Einstein (Germany)	1928	Geiger-Müller (Germany)
	• special theory of relativity		• invention of electrical radiation counter
	• photoelectric effect		
1910	Bragg (Cambridge)	1932	Chadwick (Cambridge)
	• X-rays knock particles out of the atom		• discovers neutrons
		1932	Cockroft and Walton (Cambridge)
1911	CTR Wilson (Cambridge)		• split atom using artificially-accelerated protons
	• Use of cloud chamber to identify charged particle tracks		
		1933	First cyclotron (Berkeley, California)
1911	Rutherford (Manchester)	1934	Fermi (California)
	• nucleus as source of atomic mass		• properties of slow neutrons
			• theory of beta-decay; first consistent theory of nuclear structure
1912	Bohr (Manchester)		
	• joins Rutherford		
1913	Bohr (Manchester)		
	• suggests orbital theory	1934	Joliot-Curie (France)
1913	Soddy		• discover artificial radioactivity
	• suggests existence of isotopes among radioelements	1936	Bohr and Wheeler
			• liquid drop model of nucleus
1915	Einstein (Germany)	1939	Hahn and Strassmann (Berlin)
	• general theory of relativity		• nuclear fission
1915	Moseley (Manchester/Oxford)	1944	California – first fission reaction
	• concept of atomic number		• the beginning of the study of particle physics

Figure 3.1 *The story of the atom*

Analogies often help pupils develop mental models, so you could carry out the demonstration shown in Figure 3.2. Here, the atoms in iron are compared to whole paper clips.

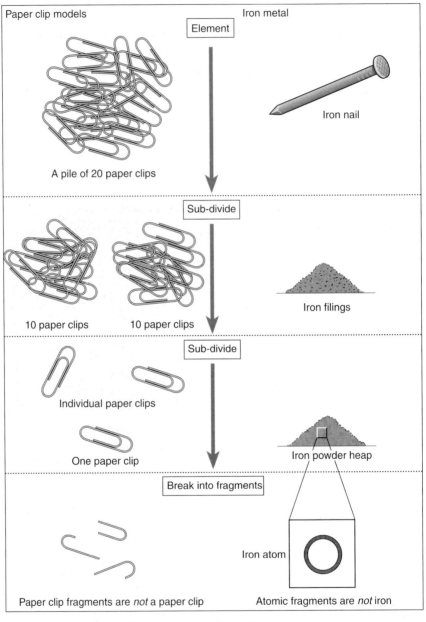

Figure 3.2 *Elements and atoms – a paper clip analogy*

The Rutherford 'solar system' model of the atom is the most useful one at Key Stage 4 which is examined by most GCSE boards and found in most GCSE textbooks. This model will be refined if pupils study chemistry at A-level and beyond.

Some pupils may have read about quarks and other sub-atomic particles and forces. This is the realm of the nuclear physicist but there are suggestions for further reading at the end of the chapter.

The CD-ROM *Elements* has an atomic theory menu with animated sequences which allows pupils to build atoms and ions, and the New Media *Atom Viewer* looks at the structure of the first 20 elements.

The paper clip analogy can be developed further by using different sizes of paper clip to help explain isotopes (see Figure 3.3).

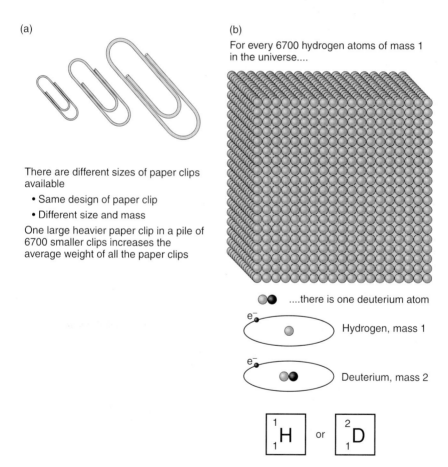

(a)

There are different sizes of paper clips available
• Same design of paper clip
• Different size and mass
One large heavier paper clip in a pile of 6700 smaller clips increases the average weight of all the paper clips

(b)

For every 6700 hydrogen atoms of mass 1 in the universe....

....there is one deuterium atom

e⁻ Hydrogen, mass 1

e⁻ Deuterium, mass 2

$^{1}_{1}H$ or $^{2}_{1}D$

Figure 3.3 *Isotopes: a) developing the paper clip analogy, b) hydrogen isotopes*

Electron arrangements

Although adding a neutron to an atomic nucleus increases its mass number and thereby generates a different isotope of the same element, adding a proton produces an entirely different element. The addition of protons increases the atomic number as well as the mass number.

The key points that pupils should be able to use to represent atomic structure are:

- that electrons move around the nucleus in orbits like the planets move around the Sun
- these orbits are not precise and predictable
- the orbits are called shells
- each shell can only hold a limited number of electrons
- the first shell can only hold two electrons
- the next shells hold eight electrons
- this rule holds true for the first 20 elements
- much of the chemistry is decided by the outer shell electrons only.

Bonding

In 1916 Gilbert Lewis realised that, with the exception of helium, all the noble gases had eight electrons in their outer shells. He linked their chemical stability to this and suggested that when elements react they try to fill their outer shells by either losing or gaining electrons.

Chemical reactions and structures are determined largely by the outer shell valence electrons. An electrovalent (ionic) bond is formed by electrostatic attraction between oppositely-charged ions. For example sodium, with one outer electron, loses this electron to achieve the atomic structure of the noble gas neon, while chlorine, with seven outer electrons, gains one electron to achieve the argon structure. So crystalline sodium chloride consists of Na^+ and Cl^- ions in a giant lattice.

Lewis went on to suggest that for some atoms it would be difficult for them to lose too many electrons because of the high ionisation energies needed to remove the electrons. Instead he suggested that they achieved noble gas status by sharing electrons, forming covalent bonds and molecules.

For example in a methane molecule, the carbon atom, with four equivalent outer shell electrons, shares with the electrons from four hydrogen atoms. The carbon atom thus has a share in eight electrons (Ne structure) whilst each hydrogen atom has a share in two electrons (He structure). Although this is an acceptable model for GCSE it is in fact a gross simplification of covalent bonding, since the actual electrons are present in molecular orbits which occupy the whole space around the five atoms of the molecule.

Co-ordinate bonding is not usually encountered at GCSE but involves the sharing of electrons, with both electrons being donated by the same atom, for example in the ammonium ion. Some textbooks refer to this as dative bonding.

The shapes of covalent compounds are influenced by electron arrangements. Bonding pairs of electrons have a tendency to be as far apart as possible in the molecule with non-bonding lone pairs of electrons having an even greater effect on the shape of the molecule. This Valence Shell Electron Pair Repulsion (VSEPR) theory is usually introduced at A-level or beyond.

The *Chemistry Set* CD-ROM has information on ionisation energies for the outer shell electrons which could be used by able pupils to predict the atoms likely to lose electrons.

You should explain to pupils that electron dot and cross diagrams are used to represent the way atoms bond together in covalent bonds. The outer shell electrons of one atom are represented by dots while those of the other atom are represented by crosses. This system is very helpful with fairly simple combinations of atoms but is of limited use when the molecules are more complex at A-level or beyond.

In summary, metals bonding to non-metals form electrovalent bonds and non-metals form covalent bonds.

Assessing pupils' learning

Ask pupils to:

- predict the type of bonds in the following: sodium bromide, methane, solder.
- produce a flick book to show what happens at electron level when sodium and chlorine react.
- draw dot and cross diagrams to show the bonding in each of the following compounds: potassium fluoride, lithium oxide, aluminium fluoride, nitrogen, ammonia, carbon tetrachloride.
- describe all the types of bonds involved in iodine solid and water at room temperature (more able pupils only).

Able pupils could produce a report which considers the concepts of atomic models, observation and indirect evidence by writing about the following:

- how is it possible to determine the characteristics of an object from indirect observations?
- how have scientists made a model of the atom?
- how are the physical properties of a substance linked to atomic structure?

ENRICHMENT AND EXTENSION ACTIVITIES

Information communication technology

CD-ROM

Elements from YITM has a good section on atomic structure.

Chemistry Set from New Media has an extensive database.

Atom Viewer from New Media has animated diagrams of the first 20 elements.

Word processing

Ask pupils to produce a report or display item on the discovery of the atom. Scan images of the scientist involved into the report.

References

Video

Atoms and Alchemists – atomic structure and radioactivity from BNFL, a teacher's pack with video and teachers' guide including over-head transparencies.

Teacher resources

Thinking about Atoms and *Thinking about Combining* – pupils' work books published by Resources in Science Education.

The Mighty Atom – information pamphlet free from BNFL.

SATIS 303 – *The Physics of Cooking*

Books

Solomon, J. *In Search of Simple Substances.* Association for Science Education

Close, P. *The Cosmic Onion*

Feynman, R. (1995) *Six Easy Pieces.* Penguin, London

Ronan, R. (1983) *The Cambridge History of Science*

2 Changing Materials

chapter **4**

Geological changes

BACKGROUND

Most pupils find this subject fascinating and you must not let your own lack of knowledge in this area put you off. Many chemistry teachers feel that this topic should not even appear in the National Curriculum, let alone the chemistry part, so consequently the subject is often presented in a very dry way. This negative approach, along with the introduction of some difficult concepts, will not help motivate pupils to learn.

The next few paragraphs will outline the key ideas that are introduced at KS4.

The Earth is approximately 4.5–4.6 billion years old. The oldest known rocks are around 4 billion years old but these are very rare. There is no fossil record of the critical period when life was first getting started.

The Earth is divided into several layers which have distinct chemical and

Table 4.1 Progression through the Key Stages

KS3 Level 3–5	KS3 Level 5–8	Foundation GCSE	Higher GCSE
• Rock classification • Rock cycle	• Process of rock formation • Weathering; erosion	• Identification of specimens	• Sequence of and evidence for the processes in the rock record • Plate tectonics

seismic properties. The crust varies considerably in thickness – it is thinner under the oceans and thicker under the continents. The information given in Table 4.2 has been derived indirectly by seismic techniques.

Table 4.2 The Earth's layers

Depth in km	Name of layer	State	Chemical composition
0–40	Crust	Solid	Quartz/SiO_2/Feldspar
40–400	Upper mantle	Fluid	Olivene, Fe/Mg; Pyroxene, Ca/Al
400–650	Transition region		
650–2700	Lower mantle	Fluid	Si, Mg, O
2700–2890	D layer		
2890–5150	Outer core	Fluid	
5150–6378	Inner core	Solid	Fe/Ni

This model of the Earth has been pieced together very recently from a number of clues from the surface and beyond. The detective work has taken place in the last 30 years. Investigations carried out in deep diamond mines in South Africa show that temperatures increase with depth and that around 3 km below the surface, temperatures reach 70°C. Geologists thought that by boring deep into the crust, they could shed light on the internal structure. The Mohole project set up in the 1960s to drill through oceanic crust into the mantle was aborted when the rocks sampled were found to be very similar to those on the surface. The evidence that geologists have used to develop the model presented has been put together like a jigsaw from a diverse range of indirect sources.

1 The first evidence that the Earth's internal structure was not straightforward was suggested by global density calculations. The density and mass of the Earth were calculated based on gravitational force. The data showed that surface rocks are not nearly dense enough to account for the total density of the planet.

2 The next clue was a rather serendipitous discovery made during the Cold War. NATO forces set up a world-wide network of seismometers which monitored nuclear explosions throughout the world. These seismic stations were also able to measure earthquakes. Geologists took advantage of the data to plot the frequency and occurrence of earthquakes. Earthquakes take place

when rocks break or slide over each other, causing vibrations which are transmitted through the Earth causing it to resonate like a giant gong. The geologists noticed that shear or S-waves were not detected on the opposite side of the globe from the epicentre of the earthquake, and that pressure or P-waves took longer to pass through to the other side than expected. Since it was known that S-waves do not pass through liquids, it was suggested that there must be a liquid core at the centre of the Earth which prevents S-waves passing through and which refracts P-waves.

3 The third piece of information was derived from surface volcanic activity. Molten magma occasionally brings up very deep rocks from the mantle onto the surface. These peridotite rocks contain minerals that can only be produced under the high pressure conditions found deep in the Earth. New global calculations of planet density which made adjustments for this mantle material were still not dense enough to account for the mass of the Earth.

4 The next piece of evidence came from outside the planet in the form of meteorites. The alignment of the planets in the solar system suggests that the Sun, planets and other rocky material have been derived from the same exploded supernova material. Occasionally some of this rocky material arrives on Earth as meteorites. This material was been found to be over 4550 million years old and to contain large amounts of iron and nickel as well as smaller amounts of other elements found in the Sun and planets. Geologists suggested that these meteorites contained materials from the early solar system. If this was the case then the Earth should have a similar composition, and it was indeed found to be an accurate assumption. When the iron was removed from the meteorite, the composition and proportions of remaining elements were similar to those found in the crust and the mantle of the Earth. This suggested that the Earth must also have much higher proportions of iron and that the most logical place to look for this was in the liquid core identified indirectly by earthquakes.

5 The last two pieces of information built on the idea that Earth has a liquid iron/nickel core. The Earth's magnetic field had been used for many centuries to navigate the globe, and records showed that magnetic north constantly changes. Geologists suggested that this movement could be explained by the presence of a liquid iron core which is in constant motion.

6 The last piece of the jigsaw is also linked to the liquid iron core. Geologists were curious as to why temperatures at the core seem to be so very high. Heat seemed to escape from the core at a fairly constant rate so in theory the Earth should be cooling down, yet the data suggested that the core seemed to remain at a fairly constant temperature. It was suggested that this may be linked to liquid iron crystallising in the centre of the core. (When substances change state on the surface there are associated energy changes. Energy is needed to change a solid to a liquid but energy is released when a liquid solidifies into a solid.) Geologists suggested that the liquid outer core may be crystallising to form the inner core and in the process releasing vast amounts of heat energy. The fluid outer core is in constant motion because of this heat exchange and as a result a magnetic field is produced by a dynamo effect in which the position of magnetic north is constantly changing direction. Geologists are, however, still unsure as to why the poles have repeatedly reversed polarity, as recorded in the mid-Atlantic ridges.

The Earth's surface is constantly changing and observers from space will have noticed a considerable difference in the distribution of land masses since the planet solidified and formed a crust. The story of plate tectonics is also an amazing account of brilliant detective work. By amassing surface evidence, geologists were able to explain, for example, why tropical animal fossils were found in Derbyshire.

Seventy-one percent of the Earth's surface is covered with water. The oceans are very important in keeping the Earth's temperature reasonably stable. Water is also responsible for most of the erosion and weathering of the Earth's continents.

Rocks are found on the rocky layer of the crust called the **lithosphere**. They are composed of minerals or aggregates of minerals. Minerals are naturally-occurring compounds with the majority having an ordered arrangement of components (crystal lattice), although a few minerals are disordered (amorphous). Rocks are sometimes glassy, for example volcanic glasses like obsidian, but usually consist of minerals that crystallised together or in sequence (metamorphic and igneous rocks), or of aggregates of detrital components (most sedimentary rocks).

The major sub-division of rocks into igneous, sedimentary and metamorphic is based on their origin and mode of formation. The process of making sedimentary rocks and erupting igneous rocks or changing existing rocks into metamorphic forms is often linked to a cyclical motion called the rock cycle.

- Igneous rocks form by crystallisation from melts and can be divided into intrusive (under the surface – magma) and extrusive (on the surface – lava) depending on whether they formed at depth (slow cooling) or at the surface (fast cooling). The slower a magma cools, the larger the crystals within the rock tend to be.
- Sedimentary rocks are formed by low-temperature processes at or near the surface of the Earth. They may consist of debris produced by weathering and erosion, animal and plant material or chemical precipitates from solution.
- Metamorphic rocks form from the first two classes (or from themselves) by

the action of heat, pressure or solutions not in equilibrium with the rock at its temperature and pressure of formation.

Unlike the other rocky planets – Mercury, Venus, and Mars – Earth's lithosphere is divided into several separate solid plates which float around independently on top of the hot mantle below. The land masses familiar to us on the surface are simply passengers on these plates. The geographical position of many land masses have altered in latitude. For example, the desert environment suggested by the rock exposures from the Devonian period can be accounted for by this movement in position. The theory that describes this is known as plate tectonics. It is characterised by two major processes: spreading and subduction.

- Spreading occurs when two plates move away from each other and new crust is created by upwelling magma from below.
- Subduction occurs when two plates collide and the edge of one sinks beneath the other and ends up being destroyed in the mantle.

The evidence for the theory has been collated very recently and began with the results of a military survey carried out in the Atlantic Ocean between 1950 and 1970. This showed that there are very high mountain ridges running the length of the ocean with some peaks equivalent in height to a subterranean Himalayas. The survey also collected magnetic field strength data from rocks on the floor on either side of the ridge. (In an igneous rock, magnetic minerals take on the orientation of the magnetic field prevailing at the time when they solidify.) The survey data showed that there is remarkable magnetic symmetry on either side of the ridge with a perfect match for polarity and field strength. It was clear that the rocks on either side of the ridge must have erupted and solidified simultaneously. Geologists suggested that new rocks are constantly being formed at the mid-Atlantic ridge with the overall effect of pushing the passenger continents further apart. This process has come to be known as sea floor spreading; it also takes place in the Pacific and Indian Oceans. Geologists postulated that as the Earth was not expanding at the same rate as it appeared to be increasing at the ocean ridges, another complimentary process must be occurring somewhere else on the globe. The earthquake belts plotted using data collected during the 1960s were found to match the plate boundaries perfectly and provided the answer.

More evidence collected along the Pacific coast and other active parts of the planet showed that the Earth's crust is actually being destroyed simultaneously in other parts of the planet.

It is now believed that convection currents in the mantle are the driving force behind the movement of plates and that mantle hot spots are often marked by volcanic activity on the surface.

KEY STAGE 3 CONCEPTS

Identifying rock specimens

Pupils are often introduced to geology at this stage through the handling and identification of rocks. This can be done by using a key or other classification method. The use of a well constructed key with good hand specimens is an effective teaching strategy. Pebble size rocks are, however, unhelpful. You can collect your own specimens or you could approach the local university department or FE college. The Earth Science Teachers' Association may be able to advise you on this. It is important that the descriptions in the key match the specimen – you may want to adapt the key to fit the specimens you have available. Clarification of crystal or grain size could be added to the key using diagrams such as those found in Figure 4.1.

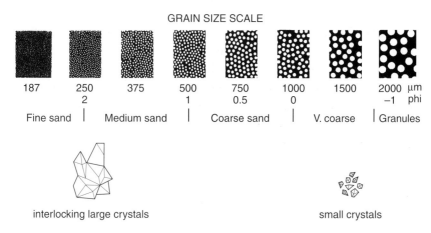

Figure 4.1 *Establishing crystal size*

To help younger pupils sort out the range of rocks available, you could use coloured pictures of the specimens to help identify the rocks. The Earth Science Teachers' Association supply photographs of polished samples or you could copy colour photographs yourself. Pupils then simply match the rock with the picture. They could then use a key like the one in Figure 4.2.

Assessing pupils' learning

Ask pupils to draw up a table with the characteristic features of each type of rock using the language of the key.

Weathering and erosion

Pupils often confuse weathering and erosion. Weathering involves the breaking down of rocks by physical methods, such as freeze/thaw action, by the action of plants and animals or by chemical attack. Erosion involves the removal of rock by rivers, the sea or the action of wind. (See Longman's *Pathways through Science: Earth and Atmosphere* module for practical activities.)

Geological time

Set the whole subject in context by introducing the time scales involved. It will help if you allow pupils the opportunity to compare geological time with a familiar physical parameter such as distance in metres or time in seconds or hours. One possible method would be to assign each pupil to a geological event or time and ask them to arrange themselves along the time line in chronological order. You could use key events such as the cooling down of the planet, the evolution of the atmosphere and the extinction of the dinosaurs, along the line. To complete the activity and to really bring home how recently man has evolved ask one pupil to move an emery board over the top of a finger nail and explain that the shavings just removed represent the length of time man has been on the Earth. Alternatively you could use the length of a day as an analogy to represent the evolution of the Earth.

Structure of the Earth

An effective demonstration which serves to illustrate how the Earth has differentiated into layers over time involves mixing approximately two parts iron filings and five parts syrup or runny honey with one part oil in a beaker. After a short time, the three substances settle into distinct layers, with the iron filings representing the core, the syrup the mantle and the oil the crust. You could also ask pupils to make a scale model of the Earth using a soft sponge ball or other material. Make slime (see Salters' *Chemistry Club Handbook*) or use silly putty to represent the mantle and show how some 'solids' can flow.

KEY STAGE 4 CONCEPTS

Rock formation and the rock cycle

Longman's *Pathways through Science: Earth and Atmosphere* module or Mike Tuke's *Earth's Science* activities suggest practical activities to show how rocks are formed. There are also fun activities using food to make 'rocks' on the BBC web page – suggestions are given for using cherries to make sedimentary squares, hazel nuts to make pudding stones or crushed biscuits to make sandstone. These are then embedded in a chocolate cementing matrix. Molten sugar with added bicarbonate can be used to simulate Hawaiian gassy lava and there is also a recipe for metamorphic marzipan. If the marzipan is coloured with food dye, the layers can be kneaded into folds and faults.

Assessing pupils' learning

Ask pupils to complete the activity in Figure 4.2.

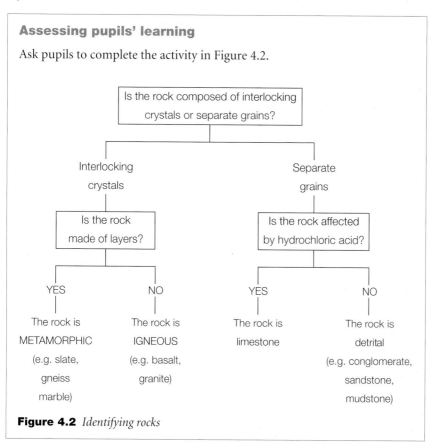

Figure 4.2 *Identifying rocks*

Plate tectonics

There are a number of excellent videos and web pages that introduce this subject really well. Mike Tuke's *Earth Science Methods* book has a range of activities to help illustrate this part of the course. The recent BBC series about the Earth has some wonderful film sequences and simulations, and the Royal Institute Lectures delivered by Dr James Jackson would provide excellent visual material and clear explanations. The Dorling Kindersley CD-ROM *Earth Quest* has simulations of earthquakes and volcanic activity.

Earthquakes

To illustrate the difference between S- and P-waves, ask pupils to stand in a line and to put their hands on the upper arm of the pupil in front. Simulate the movement of a P-wave through the mantle and liquid core by gently pushing the pupil at the back forward and sideways. The pupils in the line will move in response. To simulate the movement of S-waves, which do not pass through liquids, ask pupils to put their hands very lightly on the shoulders of the pupil in front. This time the line will not move.

ENRICHMENT AND EXTENSION ACTIVITIES

Information communication technology

Rocks and Minerals Data Files – a database from Anglia Television.

CD-ROM *Eyewitness Virtual Reality Earth Quest* from Dorling Kindersley.

Internet references
Plate tectonics
http://pubs.usgs.gov/publications/text/dynamic.html

http://www.home.earthlink.net/~dlblanc/tectonic/ptbasics.html

http://wwwneic.cr.usgs.gov/neis/plate_tectonics/rift_man.html

http://manbow.ori.u-tokyo.ac.jp/tamaki-bin/post-nuvella

Volcanoes
http://volcano.und.nodak.edu/vwdocs/vwlessons

http://www.cotf.edu/ete/modules/msese/earthsys.html

Earthquakes
http://www.hc.keio.ac.jp/earth

Chocolate rocks
http://www.bbc.co.uk/education/rocks/chocolate.html

Geology glossary
http://147.205.1581/geology/work/VFT_so_far/GeoGloss.html

Practical activities and investigations

Ask the pupils to investigate the following:
- – Which rocks are most likely to be affected by freezing and thawing of ice?
- – How do plants affect the way water runs off soil-covered slopes?
- – What is the best cementing agent for binding sand grains?

Organise field work in the local vicinity or devise a geology trail of building stone used in local shop fronts.

Visit the new Earth Galleries at the National History Museum in London.

References

Video

Royal Institution Christmas Lectures – *The story of the Earth* by Dr James Jackson, BBC Video

Science in Action – Limestone, BBC Video

Teacher resources

Clues from the Rocks – Basic Earth Science, Shell Education Service – a pack designed to support the teaching of aspects of Science to pupils aged 11 to 14.

Investigating the Quarrying Industry – Sand and Gravel, Quarry Products Association

Investigating the Quarrying Industry – Crushed Rock, Quarry Products Association

Groundwork – Introducing Earth Science from the *Earth Series 11–14*, Earth Science Teachers Association

Fisher, J. *Earth Science Field Work in the Secondary School Curriculum*, Nature Conservancy Council

Wells, A. and Tuke, M. *Rocks Around You*, Hobsons Scientific

Longman's *Pathways through Science: The Earth and Atmosphere*

Science and Technology in Society units (SATIS) are available from the Association for Science Education:

- SATIS 110
  - – Hilltop
  - – Limestone Enquiry
- SATIS 502
  - – Coal mine project

- SATIS 1001
  - Chocolate chip mining
- SATIS Atlas ASE

Books

Watt, A. *Longman Illustrated Dictionary of Geology.* Longman, Harlow

Webster, D. *Understanding the Earth.* Oliver and Boyd Ltd

Tuke, M. *Activities and Demonstrations for Teaching Earth Science.* John Murray, London

Earth Science for Secondary Teachers: An INSET Handbook NCC (1993)

Hamilton, W.R., Woolley, A.R. and Bishop, A.C. *Hamlyn Guide to Rocks, Minerals and Fossils.* Hamlyn, London

Whitehead, P. *Co-ordinated Science – The Earth.* Oxford University Press, Oxford

Atherton, M. and Robinson, R. *Study the Earth* series. Hodder & Stoughton, London

The Cambridge Encyclopaedia of Earth Sciences. Cambridge University Press, Cambridge

Suppliers

Geosupplies, 16 Station Road, Chapletown, Sheffield S30 4XM.

chapter 5

Chemical reactions

BACKGROUND

Chemical reactions are about transformations in which the combinations of atoms and ions alter, changing one substance into another. Reactions involve the addition of energy to break bonds and the release of energy when bonds are made, resulting frequently in a perceptible temperature change. During chemical reactions, electrons move from one orbit to another. The chemical industry is based on rearranging crude 'raw' materials into more useful compounds.

It is relatively rare to find a useful chemical reaction that will proceed simply by adding the reactants together. Sometimes gently heating the reactants or even just stirring or shaking them is sufficient to get things going. The use of light or an electric current will do the job in other cases. Many important industrial processes require further coercion, however, using catalysts and very high temperatures and pressures.

All reactions begin with a set of reactants which, when mixed together under suitable conditions, undergo a transformation to yield a new set of compounds, the products. This can be represented by a word equation:

$$\text{Reactant(s)} \xrightarrow{\textit{special encouragement conditions}} \text{Product(s)}$$

Some reactions appear to be a one-way process, like the burning of coal. This reaction stops when the coal runs out or the oxygen supply is removed. The reverse reaction is unlikely to occur because of the energy required to rearrange the carbon dioxide and water into the original large molecule. Other reactions appear to be reversible, for example the reaction of nitrogen and hydrogen to make ammonia. Under certain reacting conditions, the ammonia breaks down into hydrogen and nitrogen.

Some chemical reactions take place very rapidly like the explosion of TNT, while others react very slowly, for example rusting. The rate at which a reaction takes place is directly related to the ease with which the reactants can collide and rearrange into the products. Advances have been made using nanotechnology which enable measurements to be made and which even help visualise the formation of molecular bonds. This concept of collision, bond breaking and bond making is acceptable for use at Key Stage 4.

In order for a chemical reaction to take place, energy must be added to the system to activate the reactants. If a successful reaction takes place and new products are formed, energy will be released to the system. The overall energy change can be measured. This enthalpy change of the reaction is related to the change of free energy and disorder of the system. Some reactions are accompanied by perceptible temperature changes and it is a good idea to allow pupils to carry out practicals which illustrate this.

KEY STAGE 3 CONCEPTS

Many children do not question what happens when a chemical change takes place and simply accept that things just happen that way. In interviews pupils have

- described burning petrol as evaporation
- explained chemical reactions as a sort of transmutation of elements
- described decomposition reactions as simply drying or bleaching or the formation of ash.

Researchers in this field suggest that teachers should ask children in what way the new material (the product) is different from the starting material (the reactants).

It is a good idea to establish what pupils already know. One method might be to ask the pupils to sort the following changes into two categories, a change of state (a temporary change) or a chemical reaction (a permanent change):

an apple decaying, a car rusting, an ice cream melting, dry ice changing into theatrical smoke, bread being toasted, vinegar being added to a wasp sting, concrete setting, jelly setting, a plant growing and sugar being added to tea.

If the pupils are unclear of the differences between changes of state and chemical reactions, you could ask them to record the appearance before, during and after heating of the following substances: wax, ammonium chloride, zinc oxide (yellow when hot) and magnesium.

Safety Advice: Ammonium chloride will sublime. Use a small amount in a test tube loosely sealed with rocksill.

Table 5.1 Progression through the Key Stages

KS3 Level 3–5	KS3 Level 5–8	Foundation GCSE	Higher GCSE
• Permanent change	• Word equations	• Symbols equations	• Balanced equations
• Conservation of mass	• Different types of	• Endothermic	• Collision theory
• All materials and	reactions	• Exothermic	• Energy changes in
living things are made	• Useful products	• Rates of reactions	breaking and
through chemical	are made from	• Reversible reactions	making bonds
reactions	chemical reactions		
	• Energy transfer		
	accompanying		
	reactions		

S **Safety Advice:** Magnesium ribbon burns very brightly. Hold a very small piece of magnesium in a hot Bunsen flame using tongs. Warn the pupils not to look directly at the flame.

Encourage pupils to look for evidence of a chemical reaction taking place. The Royal Society of Chemistry *Microscale Chemistry* book includes a visually impressive reaction using very small quantities of lead iodide and potassium nitrate.

The decomposition reaction of copper carbonate is a safe class practical. The reaction illustrates the permanent change of green copper carbonate to black copper oxide.

- Weigh and record the mass of a large spatula of copper carbonate in a clean dry test tube.
- Insert a bung, into which a delivery tube has been fitted, into the neck of the test tube.
- Heat the tube carefully and bubble the gas evolved through limewater.
- When the copper carbonate has decomposed to copper oxide, remove the test tube from the limewater *before* you stop heating the tube to prevent 'suck back'.
- Reweigh the test tube containing copper oxide.

There will be a perceptible change in mass and the gas given off will have reacted with the limewater. Limewater is a saturated solution of calcium hydroxide which is filtered to produce a colourless liquid. This liquid will go cloudy if left in the laboratory in reagent bottles and should be freshly prepared from a stock saturated solution. Carbon dioxide reacts with the sparingly soluble calcium hydroxide to form solid calcium carbonate, which accounts for the milky appearance.

Try to help pupils to develop a mental model of what is happening by using model kits or by asking pupils to take on the role of the individual atoms involved. To do this, use five pupils – one each labelled Cu and C and three labelled O. Group the pupils together and explain that they are the $CuCO_3$ gang (reactant). Ask them to suggest a suitable rearrangement of new gangs (products) based on the evidence from their practical work.

Assessing pupils' learning

Allocate to each small group one of the following reactions:

- Heat equal amounts of copper powder and sulfur in a test tube (straight-forward).
- Weigh a few zinc granules and put them in a test tube. Add approximately 5 cm³ of 0.1 mol dm³ sulfuric acid (more difficult).
- Observe a burning candle (most difficult).

Each group should produce a theory of what reaction is taking place. The questions given below could be asked to challenge pupils' existing ideas as they carry out the practical activity.

Zinc reaction

What do you observe while the zinc and acid react? What is happening to the zinc? Which gas is given off during the reaction? Filter the zinc granules when the bubbling stops. Dry them and reweigh the granules. Is there a change in mass? Evaporate a little of the filtrate on a watchglass. Does the residue resemble zinc metal? Is the reaction between zinc and acid the same as salt dissolving in water? Use models to explain what is happening to the zinc metal and sulfuric acid.

Burning candle

Weigh a candle and watchglass and then light it. Observe the burning candle carefully, answering the following questions. How quickly would the wick burn on its own? What is actually burning when the candle is alight? What is happening to the wax? Is burning wax the same process as water evaporating? Would the candle burn if the air supply was removed? Weigh the candle on the watchglass after burning. Is there a change in mass? What purpose does the wick serve? What is happening in the bright part of the flame and in the dark part? How could you collect the gases being given off? To show the products made when a candle burns you could set up the apparatus shown in Figure 5.1.

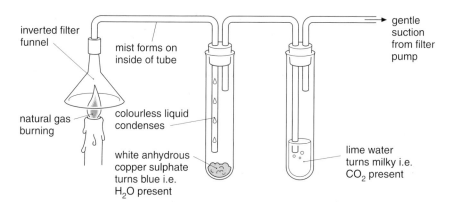

Figure 5.1 *Products of burning a candle*

You could ask pupils to make flick books to help visualise what is happening or use model kits (see Figure 5.2).

Try to help to explain what happens when the candle burns by making a reaction flick book. Make 15 blank pages before you start on card.

On pages 1 to 5 draw pictures of the reactants, in this case wax and oxygen, coming closer together until they touch.

You can draw a wax particle from a candle like this: (Wax particles are long so only a small part of one is shown here.)

You can draw oxygen particles that react with wax like this:

On pages 6 to 10 draw the particles in the reactants rearranging to form the products. Our practical evidence suggests that these are carbon dioxide and water.

You can draw carbon dioxide particles that form when wax burns like this:

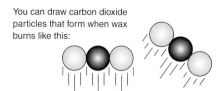

You can draw water particles that form when wax burns like this:

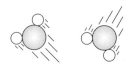

On pages 11 to 15 draw the particles in the products gradually moving further apart.

Figure 5.2 *Making a flick book to help explain what happens when a candle burns*

Types of reaction

Ask pupils to research everyday reactions like fermentation, metal displacement (see 'Metals') or neutralisation (see 'Acids'). Encourage pupils to ask what the difference is between the products and the starting materials.

Table 5.2 Everyday chemical reactions

Photosynthesis	$6CO_2 + 6H_2O \longrightarrow C_6H_{12}O_6 + 6O_2$
Respiration	Fuel (food) $+ O_2 \longrightarrow CO_2 + H_2O$
Fermentation	Sugar $\xrightarrow{\text{yeast}}$ alcohol + carbon dioxide
Thermal decomposition	$CaCO_3 \longrightarrow CaO + CO_2$
Neutralisation	Bee sting (acid) + sodium bicarbonate $\longrightarrow$ neutral
Rusting	$4Fe(s) + 3O_2(g) + 6H_2O \rightarrow 4Fe(OH)_3(s)$

KEY STAGE 4 CONCEPTS

All GCSE Chemistry textbooks have examples of practicals involving rates of reaction. At this level the collision theory of reactions is used, which states that reactions take place when particles collide. For example, in the reaction $A + B \rightarrow C + D$, in order for A and B to react they must collide successfully to make the new products $C + D$. At a very basic level this is more likely to happen if

- there are more particles of A and/or B available, i.e. the *concentration* is increased
- A and B move around more rapidly, i.e. the *temperature* is increased
- A and/or B have *more surface area* to react, i.e. one or both are crushed into smaller particles
- either A or B remains stationary/or one of the reactants is fixed to a catalyst while the other is free to move.

The main examination boards report that most of the investigations submitted in chemistry are set in this context, frequently using the reaction between marble chips and acid. Able pupils are often demotivated by this as the outcome of the investigation is obvious and not really in the spirit of investigation work. More challenging concepts and relevant contexts for practical investigations can be found in the Pupil Researcher Initiative briefs from Sheffield Hallam University.

Assessing pupils' learning

Ask pupils to set up the iodine clock reaction to change colour at a predetermined time. See *Salters' Chemistry Club Handbook* for details. You could modify and extend the instructions and ask pupils to change the concentration of each reactant in turn to decide which has the most effect on the rate of the reaction.

Energy changes in chemical reactions

Some reactions give out heat which exits from the system. These reactions are described as being **exothermic**. Other reactions take in heat from the system and are called **endothermic** reactions. There are several straightforward reactions which can easily be carried out in the laboratory to illustrate this. The temperature changes which occur with the following reactions can be measured using a data logger.

1 Ammonium nitrate in water. The ammonium nitrate dissolves in water to form ammonium nitrate solution. The overall breakdown of the lattice and the formation of hydrated ions in the solution mean that more energy is taken in than given out and so the test tube feels cold to the touch.

2 Iron filings in copper sulfate. The addition of iron filings to copper sulfate results in a displacement reaction. Iron metal changes to iron ions and copper changes from copper ions to copper metal. The overall reaction results in energy being given out to the system and so the test tube feels warm to the touch.

3 Sodium hydroxide and sulfuric acid. This is a neutralisation reaction in which hydrogen ions and hydroxide ions react to form neutral water and the salt sodium sulfate.

Print off the graphs and use these to discuss energy level diagrams. This concept is required by most syllabuses at foundation level GCSE. Use diagrams like those in Figure 5.3 to help plot the progress of the first reaction.

Photosynthesis is an example of an endothermic reaction in which energy is taken in from sunlight. This is then released in the opposite exothermic reaction, respiration.

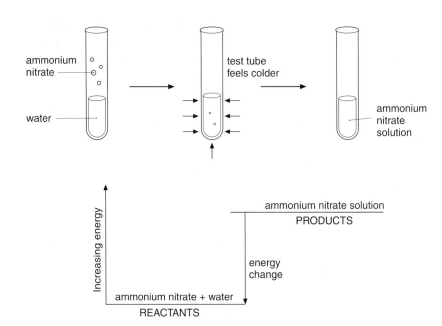

Figure 5.3 *Energy level diagrams*

> ### Assessing pupils' learning
>
> Ask the pupils to produce similar energy level diagrams for the other two reactions they have already carried out.
>
> Able pupils could be asked to write about photosynthesis and respiration or to consider the importance to evolution of the photosynthesis reaction.

Higher level GCSE candidates need to go further and consider bond energies. There are instructions in most GCSE textbooks on how to carry out practical activities to measure the amount of heat energy produced when fuels are burned. It is helpful to use model kits to help visualise the process by making models and counting the number of bonds being made and broken. Ask pupils to make one molecule of methane and two molecules of oxygen, and to complete Table 5.3 as they carry out the exercise. The reaction is

$$CH_4(g) + 2O_2(g) \rightarrow CO_2(g) + 2H_2O$$

In the table, record the number and type of the bonds broken in the reactants and remade when the products are formed. The Warwick Spreadsheet Chemistry package has an exercise using an Excel template to calculate bond energies of alcohols and alkanes.

The enthalpy change (ΔH) for the reaction is the quantity of energy transferred to and from the surroundings when the reaction is carried out in an open container. The energy change in the combustion reaction is the total energy of the products (columns c $\times$ d in Table 5.3) minus the total energy of the reactants (columns a $\times$ b in Table 5.3). Exothermic reactions have negative ΔH values and endothermic reactions have positive ΔH values.

Assessing pupils' learning

Ask pupils to use a spreadsheet to calculate the energy changes for the following reactions:

$$2H_2O(g) + O_2(g) \rightarrow 2H_2O_2$$

$$H_2(g) + I_2(g) \rightarrow 2HI(g)$$

$$N_2(g) + 3H_2(g) \rightarrow 2NH_3(g)$$

Bond energy data can be found in A-level data books.

Equilibrium reactions

Reversible reactions make products which themselves react to give back the reactants. When the two opposite reactions happen at the same rate and there is no external change, the system is said to be in dynamic equilibrium. This could be compared to a canoeist paddling upstream at the same speed as the water is flowing. The canoe's position would only change if the canoeist paddled more quickly or the river began to move more quickly.

New Media have a simulation of the Haber Process on their *Chemistry School* CD-ROM which allows the pupil to change temperature, pressure and molar masses of reactants to follow the progress of the reaction and a more advanced version is supplied by Newbyte Educational Software.

ENRICHMENT AND EXTENSION ACTIVITIES

Information communication technology

Chemistry Set CD-ROM from New Media has an extensive database, with many entries containing video sequences of chemical reactions.

New Media Multimedia *Chemistry School* has an equilibrium simulation.

Warwick Spreadsheet Package

Table 5.3 Energy changes

Bond type	Number of bonds broken (a)	Energy needed to break bond (b)	Total energy in kJ (a × b)	Bond type	Number of bonds made (c)	Energy released when bond is made (d)	Total energy in kJ (c × d)
C—H		413		H—O		464	
O=O		497		C=O		805	
		Total energy of reactants				Total energy of products	

Energy change in the combustion reaction = total energy of products − total energy of reactants

ΔH   =   (c × d)   −   (a × b)

Data logging – http://ourworld.compuserve.com/homepages/Rogerfrost/

Gas equilibrium simulation from Newbyte Educational Software.

Internet references
http://198.110.10.57/Chem/Chem1Docs/Surface Area Rate.html
http://www.intschool-leipzig.com/bailey/tutorial/rates/predict3.htm#model

Practical suggestions, demonstrations and investigations

SC1
Pupil Researcher Initiative Briefs at http://www.shu.ac.uk/schools/sci/pri
– *Catalytic Traps*, immobilising inorganic catalysts in alginate beads.
– *Lock and Key*, investigating the chemistry of enzymes.
– *A Burning Problem*, investigating the combustion of hydrocarbons.

Practical suggestions
Challenge pupils to produce a freezing mixture using citric acid and sodium
bicarbonate.

References

Videos
Scientific Eye: Series 6 Channel 4 Video

Books
Davies, K. *In Search of Solutions*. Royal Society of Chemistry

Lister, T. *Classic Chemistry Demonstrations*. Royal Society of Chemistry

Salters' Chemistry Club Handbook

Skinner, J. (1997) *Microscale Chemistry*. Royal Society of Chemistry

Wilshaw, C. and Wright, P. (1992) *Making New Materials – Science at Work*.
Longman, Harlow

Arnold, N. (1997) *Horrible Science – Chemical Chaos*. Scholastic

Jarvis, A. (1994) *Action Science – Chemical Change*. Oxford University Press,
Oxford

Holman, J. (1995) *Chemistry*. Nelson, Walton-on-Thames

Hill, G. (1990) *Making New Materials*. Hodder & Stoughton, London

chapter 6

Electrolysis

BACKGROUND

When a direct current of electricity is passed through a molten salt or an aqueous solution of an acid, base or salt, chemical reactions take place at the **electrodes**. This process is called **electrolysis**. It literally means breaking up or decomposition by electricity. Electrodes are the wires, rods or plates which make electrical contact with the electrolyte. The negative electrode is called the cathode and the positive electrode the anode. Electrodes made of graphite or platinum, which do not react chemically when in contact with electrolytes and when electricity is passed through them, are called inert electrodes.

Electrolysis is examined in virtually all of the current double and triple award GCSE examinations. It is fair to say that electrolysis is an area of the curriculum which pupils find very difficult. It brings together tricky parts of both the chemistry and physics syllabuses. When questioned, pupils ideas of electrolysis blur into garbled words about circuit boards and copper sulfate solutions. If probed further at Key Stage 4, many give rote responses and few have a real understanding of what is going on in the process. Better understanding of this concept is more likely to take place within a carefully planned teaching sequence which makes allowances for the pupils' prior learning and also builds on concepts in a logical progressive way.

KEY STAGE 3 CONCEPTS

At Key Stage 3, electrolysis is introduced as an example of a chemical reaction. Pupils often find it helpful if you break the word down and explain how it was derived. Most pupils are able to associate the *electro* part with electricity and you need only explain then that *lysis* comes from the Greek for splitting up. So electrolysis is splitting up using an electric current.

An effective demonstration of the decomposition of water into hydrogen and oxygen in a ratio of 2:1 can be carried out to illustrate this. Your school may have special equipment to do this called Hoffman apparatus, which looks like two upturned burettes joined together with electrodes fitted at the bottom. If you do not have access to this you can use upturned beakers or s-electrodes and diagrams of this can be found in most GCSE textbooks. It is important to remember that water is not a very good conductor so you will need to add approximately 10 cm^3 of 0.1 M sulfuric acid for every 100 cm^3 of water. (The sulfuric acid you find in small dispensing bottles in the chemistry laboratory is fine for this.) You can explain to the vigilant pupil that this will help the current flow more easily through the water. You can also do this as a class practical; your demonstration will reinforce the pupils' results. Some Key Stage 3 books ask pupils to explain the reaction taking place when copper sulfate is electrolysed using graphite electrodes. This is not immediately obvious to younger pupils, and I would not recommend that you spend long on this idea unless you have particularly able pupils.

KEY STAGE 4 CONCEPTS

The stage at which electrolysis is introduced to pupils will vary between schools depending on which course is being taught, but when interviewed, teachers suggested that the later on in the Key Stage 4 course they left the teaching of the topic, the more successful they were in establishing understanding. At whatever stage your scheme decides that pupils are to be introduced to electrolysis, there are a number of key ideas which need to be established beforehand. When interviewed, many year 11 pupils cited limitations in their understanding of electricity as being a stumbling block to grasping electrolysis, so it is advisable to ensure that pupils have met basic electricity first.

It might be useful to spend part of the first lesson of the sequence confirming that all pupils know that

- electrons 'flow' in a simple circuit
- a battery has a negative end and a positive end
- electrons 'flow' out of the negative end of the battery and 'flow' into the positive end
- electrons 'flow' out of the negative end of the battery at the same rate as they 'flow' into the positive end
- direct current, in which the direction of the current is constant, is needed for electrolysis
- the battery is a good source of direct current
- the battery could be considered to be an 'electron pump'.

Most published courses start by introducing the electrolysis of a molten ionic salt. Lead bromide is widely used because it has a low melting point and can be melted easily using a Bunsen burner. It is, however, quite volatile and gives off toxic vapours so the experiment should be carried out in a fume cupboard.

(S) Safety Advice: Zinc chloride is a safer alternative to lead bromide, although it should still be carried out in a fume cupboard by a teacher.

If you do not have access to a fume cupboard, you could use a video sequence of the reaction or play the video sequence from either the *Chemistry Set* or *Electrochemistry* CD-ROMs available from New Media. The demonstration shows the ionic solid being split up into the metal and non-metal parts. It is important to take time to establish the principles of electrolysis using this demonstration as an example, before continuing with aqueous solutions or industrial applications. Pupils have found it helpful to have the overall reaction broken down into stages. One suggested approach is to explain the reaction as follows:

1 Stage one: charging the electrodes

- Remind pupils that when the circuit is completed, the battery will start to 'pump' electrons around it.
- Explain that the molten salt will conduct because it has **ions** in the solution which can carry charge and so allow current to flow. It might be helpful to use a model of an **ionic lattice** to show how the ions can move once they are freed from the lattice. It is important that the pupils know that an ion is a charged **atom**.
- Explain that as soon as the circuit is complete, an electron 'traffic jam' builds up at the electrode next to the negative end of the battery. This happens because the **graphite** electrode is not as good a conductor as the copper wires connected to the battery and so there is a build up of negatively-charged electrons around it. The molten salt is not a very good conductor and adds to the 'traffic jam'.
- Introduce the correct name for the charged electrode as the **cathode**.
- Explain that the electron pump, the battery, is still pumping electrons and that the easiest place to get electrons is from the other graphite electrode next to the positive end of the battery. In the process of removing the electrons, the graphite electrode becomes positively charged.
- Introduce the correct name for the charged electrode at the positive end of the battery as the **anode**.

2 Stage two: the migration of the ions

Most pupils accept that ions migrate to the oppositely-charged electrode and can easily predict the direction of migration.

3 Stage three: the reaction of the ions at the electrode

The third stage is the one which causes pupils most problems and for some it may not be appropriate or necessary for you to introduce half equations. This stage happens at the electrodes. Two different reactions take place at the same rate at each of the electrodes.

Reaction at the cathode

1 Remind pupils that zinc or lead ions are positively charged because they have lost electrons, Zn^{2+} and Pb^{2+}, and that they are attracted towards the cathode which has an excess of electrons.

2 Explain that on arrival at the cathode the zinc or lead ions attract electrons from it. This cancels out the positive charge on the ions and leaves either zinc or lead metal.

3 This reaction for zinc is represented by the half equation:

$$Zn^{2+}(l) + 2e^- \rightarrow Zn(s)$$

or if lead is being used, then the reaction is

$$Pb^{2+}(l) + 2e^- \rightarrow Pb(s)$$

4 Point out that the ions which are attracted to the cathode are called the **cations**.

Reaction at the anode

1 Remind pupils that the chloride ions or bromide ions are negatively charged because they have extra electrons in their outer **shells** (Cl^- and Br^-).

2 Explain that when the negative ions arrive at the positively-charged anode they give up their extra electrons to the anode. The chloride or bromide ion becomes a neutral chlorine or bromine atom. As chlorine and bromine prefer to go around in pairs, however, they join up and form Cl_2 or Br_2 molecules. (See 'Covalent Bonding'.) These are gases at room temperature and can be seen forming around the anode.

3 This reaction with chloride ions is represented by the half equation

$$2Cl^-(l) \rightarrow Cl_2(g) + 2e^-$$

or if bromide ions are involved

$$2Br^-(l) \rightarrow Br_2(g) + 2e^-$$

4 Point out that ions which are attracted to the anode are called **anions**.

Assessing pupils' learning

It is important to consolidate the ideas developed so far by asking pupils to provide written evidence of their understanding.

You could ask more able pupils to explain the process in their own words with the help of some clearly labelled diagrams. It would be more challenging if you asked them to use a different ionic salt in their account.

It might be helpful to provide less able pupils with prompts or a cloze-style sheet and diagrams to label.

Going further – the electrolysis of aqueous solutions

Having firmly established the basic principles, it is easier to add some further complications. The biggest problem pupils have with the electrolysis of aqueous solutions is trying to decide which ion will react at which electrode. However if the process is introduced in a logical way, pupils are likely to find it much less problematic. It is best if pupils work in small groups of two or three to carry out the electrolysis reactions. Give each group a different solution. For each reaction pupils should

- collect any gases given off and identify them
- monitor any changes taking place in the electrolyte, including the pH
- monitor the appearance of the area around the electrodes, including any changes in pH.

Assessing pupils' learning

Able pupils could be asked to make predictions about the reaction and changes in the composition of the electrolyte.

0.1 M solutions are safe for Key Stage 4 pupils to use and the following solutions will allow all combinations of ions to be considered:

zinc chloride, sodium chloride, potassium iodide, copper sulfate, zinc sulfate, copper bromide and dilute sulfuric acid.

S **Safety Advice:** You must warn pupils that some of the products of electrolysis are quite harmful. Hydrogen is flammable, chlorine gas is toxic, and bromine and iodine solutions are harmful.

The apparatus which would be suitable can be found in most GCSE textbooks.

When the pupils have completed the electrolysis and collated their results, allow the whole class to pool their results.

> **Assessing pupils' learning**
>
> Able children could be asked to comment on patterns and arrive at the rules for the reactions at the cathode and anode.

It is important that you point out that there are now two possible cations and anions to react at each electrode because water is also present, some of which **dissociates** into H^+ and OH^-.

Cathode rules

You may need to remind pupils about the reactivity series of metals. An easy way of remembering the main metals is shown below.

Please	Potassium
Send	Sodium
Charlie's	Calcium
Monkeys	Magnesium
And	Aluminium
Zebras	Zinc
In	Iron
Lead	Lead
Cages	Copper
Most	Mercury
Securely	Silver
Guarded	Gold

Using this information it is possible to predict that when the metal ion in the electrolyte is more reactive than hydrogen and high up in the reactivity series, hydrogen gas will be discharged at the cathode. So magnesium, for example will not form at the cathode.

Anode rules

If the electrolyte contains large radicals like SO_4^{2-} then oxygen gas is given off from the OH^- in the electrolyte.

$$4OH^-(aq) - 4e^- \rightarrow 2H_2O(l) + O_2(g)$$

If there are halogen ions – chloride, bromide or iodide – in the electrolyte then chlorine, bromine or iodine will be formed at the anode.

By using the rules and considering the ions in solution, pupils find that it is fairly easy to work out what will react and what will be left in the electrolyte. For example, if sodium chloride is electrolysed the following will happen

Cations	Anions
Na	Cl^-
H⁺	*OH⁻*

Using the rules it is possible to predict that the ions in bold type will react at the electrodes and the ions in italics will be left behind in the electrolyte. Hydrogen gas will be formed at the cathode and chlorine gas at the anode. The electrolyte will change from a neutral sodium chloride solution to an alkaline sodium hydroxide solution.

Table 6.1 Summary of reactions of common electrolytes

Aqueous solution	Cathode reaction	Anode reaction	Changes to electrolyte
Zinc chloride $Zn^{2+}2Cl^-$	Grey zinc metal deposited	Chlorine gas produced which bleaches litmus	None observed
Sodium chloride Na^+Cl^-	Bubbles of hydrogen gas produced	Chlorine gas produced which bleaches litmus	Changes from neutral to alkaline
Potassium iodide K^+I^-	Bubbles of hydrogen gas produced	Iodine formed around the electrode	Changes from neutral to alkaline
Zinc sulfate $Zn^{2+}SO_4^{2-}$	Zinc metal deposited	Bubbles of oxygen gas produced	Changes from neutral to acidic
Copper sulfate $Cu^{2+}SO_4^{2-}$	Copper metal deposited	Bubbles of oxygen gas produced	Changes from dark blue to paler blue
Copper bromide $Cu^{2+}2Br^-$	Copper metal deposited	Bromine formed around the electrode	Changes from dark blue to pale blue
Dilute sulfuric acid $H^+SO_4^{2-}$	Bubbles of hydrogen gas produced	Bubbles of oxygen gas produced	Remains acidic

Check that pupils have understood these ideas and verify that they can use them by asking them to make predictions about reactions using a range of aqueous solutions.

Going further – using active electrodes

Up to this point, inert graphite electrodes have been used. The next stage will be to investigate the effects of using metal electrodes in solutions containing ions of the same metal. Copper sulfate solution and copper foil are both readily available in school laboratories and this combination is used in most textbooks.

Practical suggestion

Pupils can work in small groups using 1 mol dm^{-3} copper sulfate solution and pieces of copper foil approximately 2 cm by 5 cm. Weigh both pieces of copper foil and etch one with the symbol A+. This piece should be connected to the positive end of the battery, the anode. Etch the other piece with the symbol C− and connect this to the cathode. Set up a circuit using 6 V supply. Allow the current to flow for at least 15 minutes. Pupils should reweigh both pieces of copper foil and note any change.

S **Safety Advice:** 1 M copper sulfate solution is harmful. Provide pupils with tweezers so that they do not get copper sulfate on their hands. Pupils should wash their hands after carrying out the practical. It would be useful to collate the results from all the groups in the class and to have results collected prior to the lesson.

Assessing pupils' learning

Able pupils could be asked to explain why the anode is lighter, the cathode is heavier and the electrolyte remains dark blue after the five minutes of electrolysis.

Explain that the cathode reaction is straightforward with copper ions gaining electrons from the cathode forming copper metal.

$$\text{Cathode reaction: } Cu^{2+}(aq) + 2e^- \rightarrow Cu(s)$$

Exactly the opposite reaction happens at the anode. Copper atoms in the foil of the anode turn into copper ions. These ions go into solution and replace the

Cu^{2+} ions that are being removed at the cathode. The electrolyte remains dark blue in colour because there are still just as many copper ions in solution.

$$\text{Anode reaction: } Cu(s) \rightarrow Cu^{2+}(aq) + 2e^-$$

The *Industrial Chemistry* video from the Royal Society of Chemistry has a clip showing this process taking place on a large scale to refine impure copper metal, along with the extraction of Al and the electrolysis of brine. The *Electrochemistry* CD-ROM has a good simulation of the electroplating process.

Problem solving

Ask pupils to set up a circuit which will enable them to either copper plate the head on a 'silver' coin or to nickel plate the head on a copper coin.

Develop the idea of using active electrodes further by showing examples of objects which have been electroplated. For example EPNS written on cutlery indicates that it is in fact electro-plated nickel silver.

Assessing pupils' learning

Ask pupils to draw the circuit indicating what changes they would need to make if they wanted to gold plate an object rather than using copper plating.

Investigation opportunity

This subject lends itself readily to higher level investigation work. Pupils could be asked to investigate which factor has the greatest effect on the amount of copper deposited at the cathode. They could consider:

- concentration of electrolyte
- rate of flow of current
- shape of electrode
- distance between electrodes
- temperature of electrolyte.

Application of chemistry and nature of scientific ideas

There are two major industrial applications which are examined in this subject: the extraction of aluminium from aluminium oxide, and the electrolysis of brine.

Extraction of aluminium
The historical context Aluminium is a very common metal which pupils will have undoubtedly come into contact with. They may even have sent aluminium

fizzy drink cans for recycling. It might be of interest to them to know that the metal was not available to the ancient civilisations, or that it is the most abundant of all metals in the Earth's crust – it is more common even than iron. However, while iron has been known and prepared from its ores since prehistoric times, aluminium was not even recognised as a metal until Wohler isolated an impure sample in 1827. Henri Etienne Sainte-Claire Deville used a displacement reaction to extract enough of the metal to make Napoleon III's dinner service. In 1852 aluminium was 30 times more expensive than silver.

It was the work of Sir Humphrey Davy which led the way to aluminium being extracted. He first developed the process of electrolysis in 1807 to split up molten potassium hydroxide. His assistant, Michael Faraday, continued his work and went on to make many more discoveries about electrochemistry.

In 1866 a young American chemistry student called Charles Martin Hall realised the opportunities which would open up to anyone able to extract large amounts of aluminium metal. He set to work and discovered that aluminium oxide could be made to dissolve in a molten mineral called cryolite. Once the oxide was dissolved it was possible to extract the metal using electrolysis. In the same year the French metallurgist Paul Louis Toussaint Heroult devised essentially the same method. The Hall–Heroult method is used to this day for the extraction of aluminium. It has meant that aluminium is now very cheap to extract and is today one of the most versatile materials.

The extraction process The equipment used today operates on exactly the same principle as the other electrolysis examples we've already met. Ask pupils to look for similarities and differences between the industrial vessel and the apparatus used in the classroom turned on its side. The carbon-lined tank containing the molten electrolyte is in fact the cathode. Aluminium oxide is dissolved in cryolite (sodium hexafluoroaluminate(III), Na_3AlF_6) which melts at 955°C compared with the much higher melting point of 2072°C of pure aluminium oxide.

The anodes are made from huge blocks of graphite. High current in the region of 157 000 A at low voltages of around 4.5 V are passed through the electrolyte. The main smelting plants in Britain are located next to power stations for this reason. The Royal Society of Chemistry *Industrial* video clips show the process in operation on Anglesey.

The electrolysis of brine

The historical context The electrolysis of salt (or brine) is a major industry, producing chlorine, sodium hydroxide and hydrogen. The problem with this process is that the chlorine made at the anode mixes with the sodium hydroxide at the cathode. In 1886 Hamilton Castner set up an electrolysis cell with a mercury cathode which kept the products separate. This principle is still used today but a more environmentally-friendly membrane cell is now used.

The electrolysis process Diagrams of the basic design of the membrane cell can be found in most GCSE textbooks. The cell works continuously with sodium chloride solution flowing in one side and sodium hydroxide flowing out of the other. Hydrogen comes off continuously from the cathode and chlorine from the anode. The clever part of the process is the ion-selective membrane which only allows Na^+ ions and water through. This means that sodium hydroxide can only form in the cathode compartment, so the products cannot mix.

Assessing pupils' learning

Ask pupils to explain:

- why a power failure to the aluminium smelting plant would be a big problem
- why the graphite anodes gradually disappear during the electrolysis of aluminium
- why membrane cells are likely to be more environmentally friendly that mercury diaphragm cells
- what ions are present in the anode compartment and whether these are the same as those in the cathode department
- why the sodium hydroxide flowing out of the cathode compartment is pure and uncontaminated with chlorine.

Communication

Ask pupils to compile a glossary of new vocabulary introduced in this subject area. They could also write a letter to Sir Humphrey Davy (1778–1829) explaining that the electron was discovered later in 1897, the impact his discovery has had on our modern day lives or how his discovery actually works.

Assessing pupils' learning

Able pupils could be asked to find out how Michael Faraday arrived at such words as cathode, cation, anode and anion.

Summary of the development of concepts

	Progression	Key concept
KS3	Introduction to electrolysis	Chemical reactions
		Splitting up of compounds using electricity
KS4	Splitting up:	
	• molten salts	Electron flow and the three stages of electrolysis
	• aqueous solution	Rules for electrode reactions
	– inert electrodes	
	– active electrodes	
	Industrial application	Applying principles to more complex situations

ENRICHMENT AND EXTENSION ACTIVITIES

Information communication technology

Electrochemistry CD-ROM produced by New Media.

The Royal Society of Chemistry *Industrial Video* supplied free to all schools in the UK (1998).

Fuel cells:
http://www.i-way.co.uk/~ectechnic/electrochem
http://www.ksc.nasa.gov/shuttle/technology

chapter 7

Equations and quantitative chemistry

BACKGROUND

Many pupils find this topic difficult as it brings together the use of microscopic interpretations of reactions observed at a macroscopic level and the use of symbols. Simply telling pupils about this is not enough for some. It is important that pupils develop useful mental models which can be built on as learning progresses.

Pupils will have usually been introduced to elements (Chapter 2) and chemical reactions (Chapter 5) at Key Stage 3 before they encounter balanced equations.

Table 7.1 Progression in concepts

Concept	Cognitive development
1 Elements	Microscopic models
2 Symbols	Symbols
3 Chemical reactions	Microscopic models
4 Word equations	
5 Formulae	Symbols/microscopic models
6 Chemical equations	Ratios/proportionality
7 Moles	Symbols/microscopic models/ratios/proportionality

KEY STAGE 3 CONCEPTS

Chemical symbols

It is helpful to explain the origin of the system of symbols now in current use. It was John Dalton in 1803 who first used symbols to represent elements and compounds. He developed a series of circles to represent the elements known at the time. As there is a limited number of permutations of circle patterns and because it is difficult to remember each symbol, in 1818 Jons Jacob Berzelius, a Swedish chemist, devised the current system. He suggested using the initial letter of the element's name and when two or more elements had the same initial, adding a second letter from the body of the name. For example, hydrogen is H and calcium is Ca while chlorine is Cl. Many of the elements were given Latin names so some of the symbols are less obvious – for example, gold is Au, aurum, silver is Ag, argentum, and iron is Fe, ferrum.

Assessing pupils' learning

With the aid of a periodic table, ask each pupil to count the number of symbols it is possible to make from their name, for example LaUReN.

Writing formulae

The formula indicates the number of atoms in a molecule. For example, a molecule of hydrogen gas is made up of two atoms of hydrogen and is written as H_2. Each water molecule contains two atoms of hydrogen and one of oxygen and is written H_2O. It is important that pupils learn the rules for writing formulae before they are asked to write balanced chemical equations from word equations.

Pupils at the end of Key Stage 3 or beginning of Key Stage 4 are happy to accept the idea that each element has a fixed 'combining power', valency or oxidation state. Whether you account for this by reference to outer shell electrons will depend on the ability of the pupils and at what point in the course you introduce equations. Pupils will need to know about atomic structure if you refer to outer shell electrons. The use of Table 7.2 offers an intermediate alternative and summarises the valency information needed at this stage.

A possible method of introducing the rules for writing formulae uses sets of cards to represent some of the elements and involves working through all the possible combinations (Figure 7.1). This exercise builds on the ideas that:

- in a molecule there is no overall charge
- elements always combine together in fixed proportions
- the positive valencies are cancelled out by equal numbers of negative valencies
- there are rules for placing the subscript number
- brackets are used for large ions comprising groups of atoms which can pass unchanged through a chemical reaction.

Table 7.2 Valencies or combining powers

	One	Two	Three
Metals	Li^+, Na^+, K^+, Rb^+, Cs^+	Be^{2+}, Ca^{2+}, Mg^{2+}	Al^{3+}
	H^+	Zn^{2+}	Fe^{3+}, iron(III)
	Ag^+	Cu^{2+}, copper(II)	
	Cu^+, copper(I)	Fe^{2+}, iron(II)	
Non-metals	F^-, Cl^-, Br^-, I^-	O^{2-}, S^{2-}	
	NH_4^+, ammonium		
Radicals	OH^-, hydroxide	CO_3^{2-}, carbonate	
	NO_3^-, nitrate	SO_4^{2-}, sulfate	
	NO_2^-, nitrite		
	HCO_3^-, hydrogencarbonate		

Able pupils may be introduced to the periodic table and outer shell electrons as a method of deriving this information.

Assessing pupils' learning

Ask pupils to write down the number of atoms of each element in one molecule of the following compounds and then write formulae.

- H_2S, PCl_3, CO_2, H_2SO_4

- Iron(II) nitrate, copper(I) oxide, copper(II)oxide, sodium sulfate

---

KEY STAGE 4 CONCEPTS

Balancing chemical equations

Word equations are introduced in Chapter 5 with chemical reactions. All pupils need to use and write word equations but only some pupils will be asked to write chemical equations in examinations. You should check your syllabus for details.

Chemical equations form an international language which represents the proportions of reactants needed for a reaction to take place and the proportions of products made as a result of the reaction.

The balancing of equations can be done at two levels.

Level one

This involves balancing an equation where the formulae are already given and the pupil simply has to add numbers in front of the molecules. Work through an

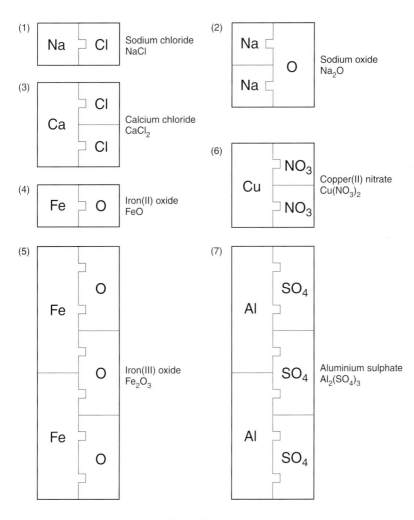

Figure 7.1 *Writing formulae – card templates*

example listing and counting the numbers of atoms on each side of the arrow. The arrow could be considered to be similar to an equals sign in a mathematical equation. In other words there must be the same number of atoms on either side. For example:

$$CH_4 + O_2 \rightarrow CO_2 + H_2O$$

1 carbon, C	1 C
4 hydrogens, H	2 H
2 oxygens, O	3 O

There are only two hydrogen atoms on the right of the equation but four on the left. Stress to the pupils that you must not change the formula at this stage. The equation is balanced by writing numbers in front of the formulae. These numbers refer to the whole formula.

It helps pupils to see why this is the case if you make models of the reactants and products. It is possible to stick small magnets to models and use any magnetic board. The equation is balanced by adding a 2 in front of both the oxygen reactant and the water product as shown.

$$CH_4 + 2O_2 \rightarrow CO_2 + 2H_2O$$

1 C	1 C
4 H	4 H
4 O	4 O

Assessing pupils' learning

Ask pupils to balance the following equations

$$Mg + HCl \rightarrow MgCl_2 + H_2$$

$$FeSO_4 + NH_4OH \rightarrow Fe(OH)_2 + (NH_4)_2SO_4$$

$$Cu + O_2 \rightarrow CuO$$

$$Na + H_2O \rightarrow NaOH + H_2$$

$$AgNO_3 + CaCl_2 \rightarrow AgCl + Ca(NO_3)_2$$

$$Fe_2O_3 + C \rightarrow Fe + CO_2$$

Level two

This level requires the pupils to write formulae before balancing the equation and means that pupils have to work with two different forms of proportionality. This can cause problems. There are four stages:

1 Write the word equation.

2 Write the formulae, checking carefully for accuracy.

3 Balance the equation. If the equation is difficult to balance, recheck the formulae.

4 Add state symbols.

Assessing pupils' learning

Ask pupils to write balanced chemical equations for:

- hydrogen + copper oxide → copper + water
- lead(II) oxide + carbon → lead + carbon dioxide

- magnesium + oxygen $\rightarrow$ magnesium oxide
- aluminium + iron(III) oxide $\rightarrow$ aluminium oxide + iron
- sodium carbonate + lead nitrate $\rightarrow$ sodium nitrite + lead carbonate
- iron(II) sulfide + sulfuric acid $\rightarrow$ iron(II) sulfate + hydrogen sulfide

For the following, pupils should decide whether the equation is correct and if it isn't, alter it so that it is. Ask them to write down the reason why they considered it to be incorrect and write notes to help a fellow pupil learn about equations.

1 $Fe + 2HCl \rightarrow Fe2Cl + H_2$

2 $2Ca + O_2 \rightarrow Ca_2O_2$

3 $2CaCO3 \rightarrow 2CaO + 2CO2$

4 $6NaOH + Al_2SO43 \rightarrow 2AlOH_3 + 3Na_2SO4$

Amount of substances

Only the most able pupils will be examined in their understanding of the mole concept. Examination boards frequently use quantitative questions to discriminate between the highest grade boundaries. You will need to decide whether you should introduce this to the pupil.

77

Explain to pupils that a balanced chemical equation tells the chemist the proportions of reactants needed for the reaction to be able to proceed. On a large plant scale the chemical engineer will want to use the correct amount of each 'ingredient', particularly if one of these is expensive. The concept is similar to the idea of following the proportions suggested in a recipe – the chef will start by weighing out the ingredients beforehand. The chemist would have difficulty weighing individual atoms so instead uses a system rather like that used in a bank. A bank cashier does not count out every coin but instead weighs the bag of coins on a set of specially-adapted scales which indicates how much each bag is worth. Chemists weigh atoms using a **mass spectrometer**. The link between the mass of an element and the number of atoms it contains is the **relative atomic mass** (A_r) of the element. It is this link which allows chemists to work out chemical formulae. The relative atomic mass scale is used to compare the masses of different atoms. The approximate relative atomic masses are listed for each element on the periodic table. There is no unit for this measurement.

Chemical quantities

Magnesium has a relative atomic mass of 24 and is twice as heavy as carbon which has an A_r of 12. A possible illustration of this involves using different sized polystyrene balls or marbles in beakers. Label the smallest beaker hydrogen and put small marbles or polystyrene balls into it, then put medium-sized balls into a medium-sized beaker labelled carbon, and finally put large balls into a large beaker labelled magnesium. Explain that the small balls represent hydrogen atoms which are the smallest atoms, that medium-sized balls represent the carbon atoms which are 12 times heavier than hydrogen atoms, and that the large balls represent magnesium atoms which are twice as heavy as carbon atoms. Explain that the beakers all contain the same number of balls of varying sizes of different masses. Point out that 1 g of hydrogen, 12 g of carbon and 24 g of magnesium all contain equal numbers of atoms because the masses are in the same ratio as the relative atomic masses. The unit that measures amount of substance in such a way that equal amounts of elements consist of equal numbers of atoms is the **mole**. Avogadro proposed this idea, suggesting that one mole of any substance contains the same number of particles. This number is called the **Avogadro number** and represents 6×10^{23} particles. Relative atomic masses (A_r), something called atomic weights, given in the periodic table tell us the amount of each element that contains this number of particles.

Using the mole in chemical calculations

There are many good books available with worked examples and questions. Three particularly useful ones are listed at the end of the chapter.

Mole calculations are based largely on ratios and proportionality. Numerate pupils will see the pattern and be able to use the mole in chemical calculations if simply shown a series of worked examples. Other pupils may need to be introduced to algorithms to help cope with the calculations. Some worked examples are set out below.

Example of a mass calculation
What mass of sulfur reacts with 1.28 g of copper to give copper sulfide.

1 Write a balanced equation $\qquad$ $Cu(s) \qquad + S(s) \rightarrow \qquad CuS(s)$

2 Decide on the mole fractions $\qquad$ 1 mole Cu $\quad$ + 1 mole S $\qquad$ 1 mole CuS

3 Calculate the number of moles of copper from the information given in the question; $1.28/64 = 0.02$ $\qquad$ [1 mole Cu = 64]

4 Substitute the mole ratios $\qquad$ 0.02 M Cu $\quad$ + 0.02 M S $\rightarrow$ 0.02 M CuS

5 Calculate actual amount in grams $\quad$ $0.02 \times 32 = 0.62$ g

Determining the formula of a simple compound
Empirical formulae of compounds are calculated by carrying out experiments

involving the measurement of the masses of the elements found in the compound. The empirical formula shows the ratio of the numbers of each type of atom in the compound. If the substance is molecular, further experiments are needed to determine the molecular formula which shows how many atoms of each type there are in a molecule. For example: what is the formula of sodium oxide if 0.69 g of sodium combine with 0.24 g of oxygen?

	Sodium	Oxygen
	0.69 g	0.24 g
Mass of one mole	23 g	16 g
No. of moles available	0.69/23 = 0.030	0.24/16 = 0.015
Simplest ratio	2 : 1	
Empirical formula	Na_2O	

If will help pupils if they remember that:

$$\frac{\text{Mass (g)}}{\text{Formula mass}} = \text{Number of moles}$$

Worked examples for other calculations are found at the Phoenix College website given below.

ENRICHMENT AND EXTENSION ACTIVITIES

Information and communications technology

Internet tutorial web page – http://dbhs.wwvsd.K12.ca.us/meaning-ofEquations.html
GCSE CD-ROM, *Aircom GCSE Science*
Pheonix College – http://chemlab.pc.maricopa.edu/labbooks/

References

Books
Brown, K. (1986) *Moles, a Survival Guide for GCSE Chemistry.* Cambridge University Press, Cambridge

Ramsden, E.N. (1993) *Calculations for GCSE Chemistry NC Edition.* Stanley Thornes, Cheltenham

Pathways through Science: Making Materials Skills Sheets (1993). Longman, Harlow

chapter 8

Useful products from oil

BACKGROUND

This section of the National Curriculum Programme of Study is included in both the single and double award syllabuses because organic chemistry is a very important part of the chemical industry and has many everyday applications. Crude oil is the main raw material used by the pharmaceutical and petrochemical industry, and is the liquid part of petroleum. Petroleum is a naturally-occurring mixture of hydrocarbon compounds. Other examples of hydrocarbon compounds are solid asphalt and natural gas.

KEY STAGE 3 CONCEPTS

At KS3 pupils are introduced to the separation of mixtures by fractional distillation. Crude oil is often used as an everyday example of a mixture separated by this method. The Key Stage 3 Programme of Study requires that pupils also learn about the formation of oil deposits and here it would be advisable to liaise with the Geography department to avoid duplication. There are many free resources on offer which will help illustrate the subject, for example the Institute of Petroleum has set up a very useful web site from which up-to-date information can be downloaded.

Pupils could write a report or produce a pamphlet addressed to a younger audience outlining how crude oil is formed. To encourage pupils to use this downloaded information appropriately, make the assessment criteria explicit when giving out the task, and reward accomplishment in the key skills as well as in knowledge-based work. Dissuade pupils from simply submitting chunks of this text.

KEY STAGE 4 CONCEPTS

At Key Stage 4 the important concepts are

- the refining of crude oil
- the nature of the fractions produced
- the chemical processing involved in making the fractions more useful in everyday life.

Refining crude oil

Pupils should know that crude oil is a mixture of compounds – **hydrocarbons** – which contain hydrogen and carbon only. These hydrocarbon chain molecules have different lengths and therefore different boiling points and can be separated by fractional distillation.

There are a number of videos available which show this process on an industrial scale, and the fractional distillation of a substitute 'crude oil' mixture can be simulated in the laboratory in a well-ventilated room. Crude oil should not be used as it contains benzene, however you can make up a suitable mixture which will split into fractions on heating.

Table 8.1 Substitute mixture for the laboratory fractional distillation of crude oil

Crude oil mixture	Volume (cm³)
Liquid paraffin	55
Paraffin oil (kerosene)	20
White spirit	11
Petroleum ether (100–120°C)	4
(80–100°C)	4
(60–80°C)	4
One spatula of powdered oil paint or charcoal	

Safety Advice: You should check your school safety regulations and carry out a risk assessment about when it may be possible to allow pupils to perform the separation with close teacher supervision. An alternative is to show examples of the fractions obtained from oil. BP supply a boxed set of crude oil fractions which help illustrate this. The proportions of each of these fractions varies between regions and oil wells. Fuel oil is the fraction which is in most demand.

Pupils should have a clear idea about the temperature ranges in the fractionating column and at which temperature each fraction is extracted. The Institute of Petroleum's free publication, *Fossils into Fuels*, provides good clear diagrams of the fractional distillation tower, including explanations of how the process uses bubble caps to improve separation. A black and white version of this is available on their web site.

Assessing pupils' learning

Longman's *Pathways through Science: Making Materials* provides a photocopiable sheet which asks pupils to cut and paste a jumbled diagram of a fractional distillation column in the correct position. Alternatively, an outline of a distillation tower could be labelled and notes made on the physical properties and uses of each fraction.

The nature of the fractions produced

- At Key Stage 4, pupils should know how to name simple organic molecules and be aware of the differences between saturated and unsaturated compounds.
- Most syllabuses require pupils to be able to carry out the test for unsaturation by reacting compounds with bromine water.
- Higher level papers require pupils to recognise geometric isomers.
- Pupils also need to know that long-chain hydrocarbons which are less economically useful are broken down into smaller chains by the cracking reaction.

Naming organic molecules

The naming of organic molecules sometimes causes problems which can be reduced if the international naming system is introduced as a logical process, together with the use of model kits, which will help pupils visualise the patterns.

You could start with the simplest organic compounds from the homologous series of **alkanes**. Ask pupils to make models of methane and ethane and explain that both molecules are from the same family group, or homologous series, called alkanes. Each member of the family has the same general type of molecule, a chain of carbon atoms, with hydrogen atoms attached to the side positions. This is distinctive to this family of hydrocarbons and is sometimes called the **functional group**. Each member of the family has a different chain length and therefore a different name but the names all end in -**ane**. Since carbon can form chains of varying length, a system has been developed in which the prefix of the name indicates the number of carbons in the chain. For example, methane has one carbon atom and ethane has two. In summary, therefore, when naming hydrocarbons, the prefix indicates the chain length while the suffix denotes the family group. Tables 8.2 and 8.3 list some of the most common prefixes and suffixes.

Double award syllabuses require pupils to name alkanes and alkenes only, but some chemistry syllabuses may require pupils to name alcohols and organic acids as well.

Compare models of ethane and ethene and ask pupils to describe the differences.

They should point out that alkenes contain fewer hydrogen atoms than alkanes or that the carbon atoms have fewer bonds available to join to hydrogen because of the C=C double bond. As a result of this, alkenes are said to be **unsaturated** and alkanes **saturated**. You could show them how to distinguish between the molecules by using bromine water which reacts with the double bond in the alkene to give a colourless compound. This will be discussed later in the chapter.

83

Table 8.2 Prefixes used to identify chain length in family groups.

Number of carbon atoms	Prefix in the name
1	meth-
2	eth-
3	prop-
4	but-
5	pent-
6	hex-
7	hept-
8	oct-

Table 8.3 Suffixes used to denote homologous series

Functional group	Name ending	Homologous series
C—C	-ane	alkanes
C=C	-ene	alkene
C—O—H	-ol	alcohol
CO_2H	acid	acids

Isomers

When carbon chains are at least four carbon atoms long, they are able to form branched molecules as well as straight-chain molecules. Ask pupils to build two different structures using model kits, for a molecule containing four carbon atoms and 10 hydrogen atoms. The use of models helps pupils to visualise the spatial arrangements and emphasises the importance of the shape of organic molecules which is not obvious when using 2-D diagrams in textbooks. The two possible arrangements are shown in Figure 8.1.

Isomers are compounds with the same molecular formula but different structural or graphical formulae. The straight chain molecule is butane and the branched molecule, 2-methylpropane, is a different compound with different physical properties.

The more able pupils should know how to name compounds with branched chains. Explain the rules for doing this using the branched isomer of butane, shown in Figure 8.1, as an example.

1 The longest chain is three carbons long so the first part of the name is prop-.

2 The molecule has alkane functional groups, C—C/C—H groups so the chain is propane.

3 It has a methyl group branched on the second carbon in the chain, so its full name is 2-methylpropane.

Some syllabuses require more able pupils to use the empirical formula – the lowest ratio of atoms in a molecule – to work out the molecular formula of the compound. See 'Quantitative Chemistry'.

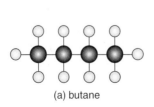
(a) butane

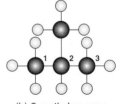
(b) 2-methylpropane

Figure 8.1 *Isomers of C_4H_{10}*

Chemical processing

Crude oil is the raw material from which a very wide range of products are made. Three important chemical reactions starting with crude oil are introduced at Key Stage 4 – catalytic cracking, polymerisation and fermentation. The first two are discussed in detail below.

Cracking

Petrol, ethene and short-chain fractions are more valuable in the market place than some of the longer-chain products from crude oil and are therefore more in demand. In general there is too much of the higher boiling point, long-chain hydrocarbon fractions and too little of the lower boiling point, short-chain fractions. If an oil company is to make a profit, it needs to find ways of using all the products of crude oil distillation.

Oil companies **crack** longer-chain hydrocarbons by heating them under pressure and at fairly high temperatures, normally with a catalyst and sometimes in an atmosphere of hydrogen. This breaks the long chains into shorter more useful ones. There are a number of videos which show this process on an industrial scale, or you could illustrate it on a smaller scale in the laboratory by cracking liquid paraffin, petroleum jelly or decane. The procedure could be carried out by the pupils in a well-ventilated room with close teacher supervision. The apparatus required can be found in most GCSE textbooks or CLEAPPS Hazcards.

S **Safety Advice:** The gas produced is highly flammable. Once enough gas has been collected, disconnect the apparatus to avoid suck-back.

The products of the cracking reaction can be tested for unsaturation using bromine water. Unsaturated compounds – those containing a $C=C$ double bond – will decolourise the bromine liquid. Use very dilute solutions of bromine, otherwise the colour change may not be observed. Dispense about 1 cm³ of the test solution, cracked gas or liquid alkene (cyclohexene or oct-1-ene work well) in a fume cupboard or a well ventilated room. Shake the test tube and observe the changes.

S **Safety Advice:** Very small amounts of toxic 1,2-dibromoethane are formed when ethene is used as the starting material.

Assessing pupils' learning

A suitable problem-solving activity which brings together knowledge of the cracking process, the test for unsaturation and a range of practical skills involves investigating a fictitious leak in a cracking plant. A schematic diagram of a cracking plant is given to the pupil as an introduction to the industrial process.

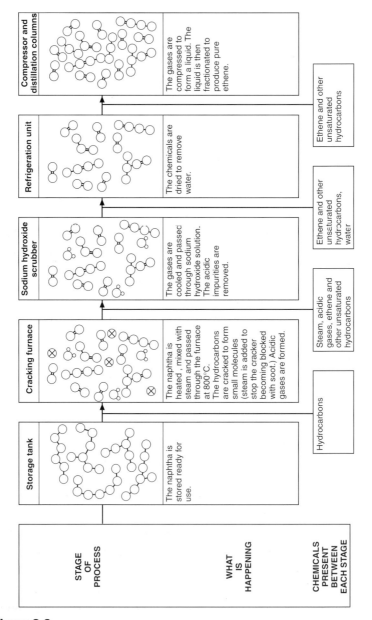

Figure 8.2

Pupils are then asked to locate the leak based on the schematic diagram and their analysis of an oily sample (water contaminated with small amounts of cyclohexene) collected from drains in the plant. They could then write a report of their investigation or transform the schematic plant diagram information into a flow diagram. This activity is based on one suggested in a pack called *Chemistry in Action* which may now be out of print. There may well be a copy on the shelf in your chemistry prep room along with the video produced by Granada TV some years ago.

Polymers

Polymers can be divided into three groups depending on what they are made from.

- **Natural polymers** come from animal and vegetable sources. They include cotton and wood, made from condensation polymers, cellulose, and wool or silk made from protein.
- **Synthetic polymers** are usually made from petroleum or coal. They include the vast range of plastic materials widely available.
- **Semi-synthetic polymers** are based on natural materials which are chemically modified. They include vulcanised rubber, rayon (made from cellulose) and casein (made from proteins in milk).

One possible way of introducing this subject would be to encourage pupils to survey the range of polymers that they come into contact with in one day. The international symbols, which will help identify the plastic material, are nearly always stamped on the item (see Figure 8.3).

Pupils need to know that polymers are long-chain molecules made by joining together large numbers of small molecules called **monomers**. The two types of reaction tested in GCSE examinations are:

1 Addition polymerisation

Use models to simulate an addition polymerisation reaction. Ask each pupil to make an ethene molecule and then to stand in a line along the side of the room. Initiate the 'reaction' by asking the first person in the line to break the double bond in their model, which will start a chain reaction in the line to produce a long chain of polythene. Some pupils will invariably try to stop the process by adding branches but this will provide you with the opportunity to discuss potential problems encountered by plastic manufacturers. Point out that this is an example of an addition polymerisation reaction in which poly(ethene), more commonly known as polythene, is the polymer produced from ethene monomers.

PET or **PETE** (polyethylene terephthalate)

Most commonly recycled plastic, used to make two-litre fizzy drink bottles and plastic liquor bottles. Recycled into many products such as bottles for cleaning products and non-food items, egg cartons, and fibres (carpet, t-shirts, fleece, etc.).

 HDPE (high density polyethylene)

Commonly used to make milk and juice bottles. Recycled into many products such as lumber substitutes, base cups for soft drink bottles, flower-pots, pipes, toys, buckets and drums, traffic barrier cones, bottle carriers and rubbish bins.

 V or **PVC** (polyvinyl chloride)

Used to make flooring, shower curtains, house sidings, garden hoses and many other products. Not currently recycled.

 LDPE (low density polyethylene)

Used to make cellophane wrap, disposable nappy liners and squeeze bottles. Not commonly recycled.

 PP (polypropylene)

Used to make packaging pipes, tubes and long underwear. Not commonly recycled.

 PS (polystyrene)

You may know this as 'styrofoam'. Used to make coffee cups, take-away food packaging and egg cartons. Recycled in some areas and made into the same type of products, insulation, plastic 'wood' and hard plastic pens.

 Others (all other plastic resins or a mixture of resins)

Not commonly recycled.

Figure 8.3 *International recycling symbols (plastic)*

2 Condensation polymerisation

It is possible to carry out a condensation polymerisation reaction in the laboratory. Making nylon is very effective as a teacher demonstration or, if very small volumes of the reagents are used and there is very close teacher supervision, it is possible to allow pupils to make the polymer.

In this reaction two different monomers join together to make the long nylon chain and water molecules, which is why the reaction is called condensation polymerisation (see Figure 8.4).

Most pupils are happy to accept the representation of the molecules in a schematic form but you could give the names and structural formulae to more able pupils.

The templates in Figure 8.5 can be used to cut out coloured acetate sheets for use on an overhead projector to show how the reaction proceeds.

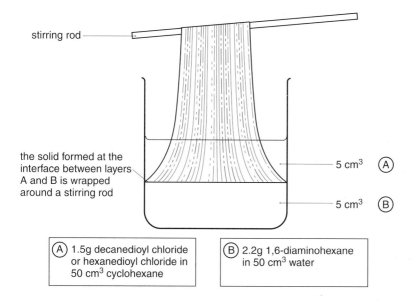

Figure 8.4 *Making nylon*

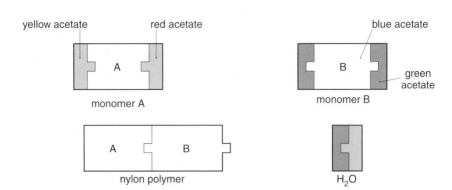

Figure 8.5 *Templates for use with the overhead projector*

Assessing pupils' learning

Pupils could be asked to draw a sequence of diagrams to show the polymerisation reaction and to name the polymer formed when the monomers in Table 8.4 are used.

Pupils with well developed ICT skills could produce an animated sequence to show both types of polymerisation reaction.

Table 8.4 Polymerisation reactions

Polymer	Monomer	Molecular formula of monomer	Structural formula of monomer	Structural formula of polymer
Poly(ethene)	Ethene	C_2H_4		
Poly(chloroethene) also called poly vinyl chloride	Chloroethene	C_2H_3Cl		
Poly(phenylethene) also called polystyrene	Phenylethene	$C_6H_5-C_2H_3$		
Poly(methyl methacrylate) also called perspex	Methyl methacrylate	$CH_3-CH_3.COOCH_3$		

3 Structure and function in polymers

Many of the man-made polymer materials mimic the structure of naturally-occurring polymers. Pupils will meet photosynthesis in their biology lessons or as an example of an endothermic reaction elsewhere in their chemistry courses. One of the products of this reaction is glucose. Glucose is the monomer from which polysaccharide molecules are made by condensation polymerisation. The plant can change the function of the polymer in response to external stimuli by altering the structure of the polymer. For example, when a plant is at the start of the growing season it needs polymers that are strong so it produces 'scaffolding' molecules – cellulose. Later in the growing season the plant may want to store sugar molecules and so starch, another natural polymer made from glucose, will be made and stored in a root tuber, in the case of the potato.

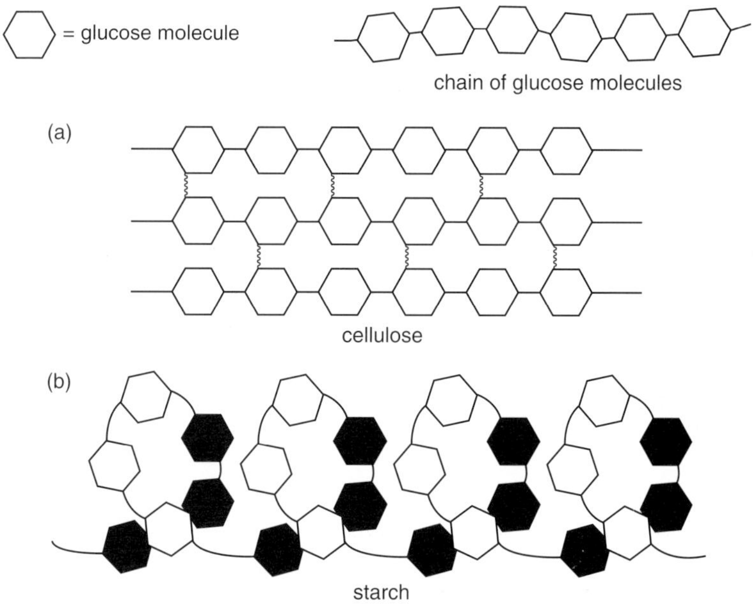

Figure 8.6 *Structure of cellulose and starch*

Pupils could identify polymers by using different density liquids. There are detailed practical instructions in the *CIEC Recycling City Pack*. Burning tests should be carried out on a very small scale and in a fume cupboard (see Figure 8.7).

S **Safety Advice:** Many products of pyrolysis reactions are very toxic. Avoid carrying out the test on PVC. Consult your LEA safety advice.

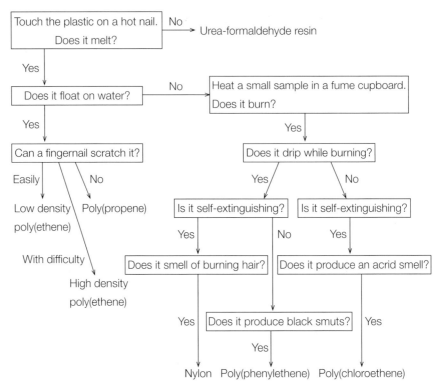

Figure 8.7 *Identifying plastics*

Thermoplastics and thermosets

Polymers (or plastics) can be classified as either *thermosetting* or *thermosoftening* plastics according to how they change on heating. For example, polythene and nylon soften when heated and harden when cooled. These are called thermosoftening plastics. Other polymers soften when heated but then harden permanently and the material will not soften again. These are called thermosetting plastics. In a thermosetting plastic, strong covalent bonds form cross links between the chains so it is not possible for the chains to become free without breaking down the whole structure.

ENRICHMENT AND EXTENSION ACTIVITIES

Information communication technology

Institute of Petroleum, http://www.petroleum.co.uk/
Chevron, http://www.chevron.com

Practical activities and investigations

Design and carry out an experiment to find out whether samples of cooking oils, lard, margarine and butter contain saturates, unsaturates or polyunsaturates.

References

Videos

Refining, which describes the chemistry of oil refining, is available on free loan from BP.

Petrochemicals, which describes cracking, polymerisation and polymers, is available from Shell (UK).

Teacher resources

Chemical Industry Education Centre (CIEC) resources:
- Product design, technology/science units for 14- to 16-year-old pupils:
  - Polyurethane in sports shoes
  - Polymers in sleeping bags
  - PET in pop bottles
- *POLYPACK*, a polymer investigations kit which includes samples of different polymers and teachers notes.
- *Recycling City*, the uses and recycling of polymers including practical instructions for the identification of plastics using density.
- *Let Sleeping Bags Lie*, suitable for KS3, looks at the choice of materials for sleeping bags.
- *About Plastics*, an information pack for 14+ from Shell Education, includes full colour booklets.
- *Platform Pack* for 14–16 science, including information on polymers, monomers, fractional distillation, polymerisation, properties of plastics and recycling, free from the Association of Plastic Manufacturers in Europe.

Books

McDuell, B. (1990) *Science Now: Plastics a Plenty*. Stanley Thornes, Cheltenham

Pathways through Science: Making Materials (1993) Longman, Harlow

93

chapter 9

Useful products from metals

BACKGROUND

Metals consist of giant covalent structures in which each atom contributes one or more of its valence electrons to the formation of an omnidirectional delocalised covalent bond that extends throughout the structure. This is often referred to as a 'sea of electrons' in GCSE textbooks.

Metals have a regular crystalline structure. You can see crystals of lead form if you immerse a strip of zinc in a solution of lead(II) nitrate (see 'Atomic Structure and Bonding'). The main properties of metals are:

- they have a *lustrous appearance* (shiny) because of the reflection of light from the regular arrangement of atoms.
- they are also *good conductors of heat* because of this structure.
- sound waves are able to travel through the lattice and so metals are *sonorous* – they ring when struck.
- within the metallic structure, the atoms give up some of their outer electrons and become positive ions while the electrons released move freely in the solid as a whole. They are, therefore, very *good conductors of electricity.*
- they are *ductile* – able to be pulled into wires – and malleable – able to be hammered into sheets. It is possible using models to show how layers of copper atoms, for example, can slip over each other.

KEY STAGE 3 CONCEPTS

Pupils' preconceptions of what metals look like are challenged by the appearances of alkali metals. You may like to link up a circuit to show that they do conduct electricity.

Pupils also find the displacement of metals difficult to understand, particularly in a solution of their salts. The metals displaced in this way do not resemble the more familiar foil samples.

Properties of metals

Establish what pupils have learned already, perhaps by asking them to draw a concept map. You could set the subject in context by considering how important metals are in everyday life by asking pupils to survey the number of metals encountered over a 24-hour period and suggest which metallic property is important in this role. For example, copper metal inside a plug is used because it is ductile and can be pulled into a wire, is a good conductor and is reasonably cheap.

Reactions of metals

Most schemes of work have practical tasks which will illustrate reactions of metals with water, air and acids.

- *Starting Science* has a series of practical tasks in which samples of metal are left exposed to the air for a long period, reacted with water and reacted with acids.
- Nuffield *Science for Key Stage 3* (Y9) also has a good range of activities.

Pupils should be able to handle most metals. Table 9.1 summarises the reactions with acids.

Safety Advice: Pupils should not handle alkali metals so you may like to demonstrate the reactions of these metals with water and oxygen. Add universal indicator to the water before dropping in the alkali metal – it will change from green (neutral) to purple (strong alkali).

When handling alkali metals you must wear eye protection and use a safety screen. It is best if you use prepared pieces of metal with sides no longer than 4 mm. *Never* attempt to constrain the sodium; allow it to roam freely on a large surface of water.

If you do not feel confident about carrying out the demonstration, you could show pupils the video sequence on the *Chemistry School* CD-ROM. The *Chemistry Set* CD-ROM also shows the reactions of rubidium and caesium.

Reactivity series
You could ask pupils to work out an order of reactivity of metals from the practical work they have carried out. There is a good summary of the reactions of metals in *Chemistry Revise through Diagrams*.

Table 9.1 Summary of reactions of metals with acid

Metal	Safety advice
Ca and Li	Very exothermic. Use only a small grain of Ca or Li. Teacher demonstration only at KS3.
Mg	Very fast and exothermic. Be careful that the acid does not rise out of the test tube. Use ~2 cm ribbon.
Al	Very slow at first because of the Al_2O_3 layer, then quite vigorous.
Zn	Reaction is slow with granulated zinc; if making H_2 gas, a few granules of copper(II) sulfate act as a catalyst.
Fe	Solution may be gently warmed.
Pb	Does not react with H_2SO_4; slight reaction with HCl. Pupils should wear gloves.

DO NOT react Na, K or other reactive metals with acids

An aide to help pupils remember the reactivity series of metals is given in Chapter 6. Alternatively you could ask pupils to produce a set of different cards with the information suggested in Figure 9.1. They could play rummy, snap or Top Trumps with different permutations of the properties of each metal.

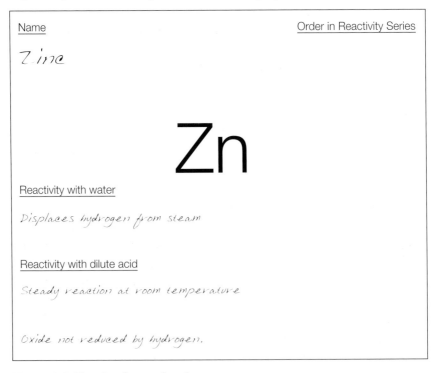

Figure 9.1 *Template for metal card*

Assessing pupils' learning

We can compare the reactivity of metals by testing all of them against the same element. Some of the results are shown in Table 9.2. The reactivity of each metal has been tested against hydrogen. Imagine that the metals are football teams all playing Hydrogen Town.

Table 9.2 Metal activity league

MAGNESIUM ATHLETIC	7	HYDROGEN TOWN	0
ZINC WANDERERS	4	HYDROGEN TOWN	1
ALUMINIUM CITY	5	HYDROGEN TOWN	1
PLATINUM ROVERS	1	HYDROGEN TOWN	5
SODIUM UNITED	9	HYDROGEN TOWN	0
LEAD ORIENT	4	HYDROGEN TOWN	2
COPPER COUNTY	0	HYDROGEN TOWN	1

Now put the metals in order of reactivity. The most reactive metal goes at the top, the least reactive at the bottom.

More able pupils could consider the following:

Cr is below Zn but above Fe in the reactivity series. Use the series to help you predict what reaction, if any, you would see if Cr is a) heated in air, b) added to cold water, c) added to hydrochloric acid.

Displacement reactions

The nucleus of every atom is surrounded by electrons and the electron density in a pure element is taken as normal or standard. However, an electrode of any element in contact with a solution of its ions exhibits an **electrode potential**. For example, calcium metal ions are surrounded by delocalised electrons and in solution the ions are surrounded by lone pairs of electrons from water molecules. The electrode potential is the difference between the two, and the magnitude of this potential depends on both the element and the concentration of the solution. The change in electron density around atoms during oxidation and reduction means that there is a difference in electrical potential between the oxidised and reduced forms of the element.

This potential difference is measured under standard temperature and pressure conditions and the electrode potential values presented in data books are relative values – they are always compared to the hydrogen electrode potential. The electrochemical series and the reactivity series found in Chapter 6 are virtually the same. You can suggest that pupils use this reactivity series to predict which metal will be displaced. They should use the principle that a more reactive metal displaces a less reactive metal from its compound. This can be demonstrated effectively by cutting a piece of copper foil into an interesting shape like a Christmas tree and standing the foil shape in a small beaker. When 0.2 mol dm^{-3} silver nitrate solution is poured carefully into the beaker, a displacement reaction is initiated with copper ions going into solution and silver metal being deposited in its place.

(S) Safety Advice: Silver nitrate is an irritant at this concentration and it is advisable to wear gloves to avoid staining your hands.

A spectacular displacement reaction is used to weld railway sleepers in situ; the details are given below. The **thermit reaction** involves the displacement of iron from iron oxide by aluminium to form aluminium oxide and iron metal. It is very impressive but if you are not confident or do not have the necessary equipment, you could show a video sequence or the extract from the AI page on the *Chemistry Set* CD-ROM.

(S) Safety Advice: Carry out a risk assessment before attempting this demonstration. Depending on how confident you are with the group, there are two ways of carrying it out.

1 Safe tried and tested Fill the mixture into a piece of folded filter paper and place this in a sand tray or fire bucket.

2 More risqué Use a new ceramic flower pot and clamp it on a stand about 1 metre above a sand tray or fire bucket. Surround the pot and tray with safety screens or carry out the demonstration in the fume cupboard. The molten iron produced will flow from the bottom of the pot.

Thermit reaction – teacher demonstration

1 Weigh out the mixture:16 g iron(II) oxide, Fe_2O_3, and 4 g aluminium powder. This mixture is relatively stable and needs a vigorous starter mixture to set it off.

2 Mix the reactants very well and pour into the chosen reaction vessel.

3 To get the reaction started, make a 4 cm deep cavity in the thermit mix and pour in 1.7 g barium peroxide (oxidising and harmful) and 0.2 g magnesium powder (highly flammable).

4 Set up a safety screen and wear eye protection.

5 Finally twist a little magnesium ribbon and push it into the starter mixture and down into the reactants – this is the fuse. Light the fuse and move to a safe distance.

It is important that you explain exactly what has gone on. There is a danger with chemistry demonstrations that the concept which is to be learned is lost in the bangs and smells. Depending on how politically correct you want to be you could use pupils to represent the elements in the reaction by choosing two boys to represent aluminium and iron, and one girl to represent oxygen. The equation for the reaction is:

$$Fe_2O_3 + 2Al \rightarrow 2Fe + Al_2O_3$$

You can suggest that aluminium is the 'bigger bully', or is more attractive to 'Miss Oxygen', and that when the reaction is initiated, oxygen combines with aluminium leaving iron on its own.

Having looked at the displacement of oxygen it is then possible to consider displacement of metals from solutions of metal salts.

> **Assessing pupils' learning**
>
> Ask pupils to make predictions and test them for the combinations of magnesium, iron, copper metal and solutions of magnesium, iron, copper and silver salts. Point out that this time the more reactive metal will displace the less reactive metal ions out of the solution.

Extracting metals from ores

Minerals are naturally-occurring compounds and ores are useful minerals found in large enough concentrations to make it economically viable to extract them. Table 9.3 summarises some common ores.

It is possible for the class to extract lead metal from lead oxide. This should be carried out in a well-ventilated room. Explain that the carbon has 'grabbed' the

Table 9.3 Some common ores

Metal	Name of ore	What's in the ore
Al	Bauxite	Al_2O_3
Cu	Copper pyrite	$CuFeS_2$
	Malachite	$CuCO_3$
Fe	Haematite	Fe_2O_3
Na	Rock salt	$NaCl$
Sn	Cassiterite	SnO_2
Zn	Zinc blende	ZnS

oxygen from the lead. You could use pupils as 'Miss Oxygen' or 'Mr Carbon' or 'Mr Lead'. Alternatively you could help pupils visualise this reaction by drawing cartoon particle pictures of metal extraction or using model kits.

The equation for the reaction is

$$2PbO + C \rightarrow CO_2 + 2Pb$$

Assessing pupils' learning

Ask pupils to:

- write equations or particle pictures for reductions (removal of oxygen) by carbon of copper oxide, zinc oxide and tin oxide.
- link the date of discovery of the metal with its position on the reactivity series – gold is easy to extract because it is very unreactive.
- find out which advance in technology has enabled the metal to be extracted.

Table 9.4 Date of discovery of metals

	Date	Metal discovered	Technological advance
	1900AD	Aluminium	
	1800AD	Sodium, potassium,	Electricity
		magnesium, calcium, strontium	
	1700AD	Nickel	Displacement reactions
	1500AD		
	1000AD		
	500AD		
	300AD	Mercury distilled from ore	
Iron Age	0		
	50BC	Brass made	Alloying
Bronze Age	500BC		
	1000BC	Bronze	
	1000BC	Iron	Reduction of Fe_2O_3
	1500BC		
		Possibly lead/zinc reduced	
		accidentally by carbon	
Stone Age			
		Copper, tin, silver	Fire
	3000BC		
		Gold, silver	

The extraction of aluminium is considered in the chapter on electrolysis (Chapter 6). Many textbooks look at the extraction of iron from the blast furnace. The ISCOR Ltd website has an animation of the process and the Royal Society of Chemistry industrial video also shows the process clearly. It would be helpful if you could explain this process as a logical sequence of reactions with pupils being involved in an active process whereby they sort this complex multistep reaction process out for themselves. Simply labelling a diagram of the blast furnace will encourage shallow rote learning. The key reactions are:

1 The coke burns.

$$C(s) + O_2(g) \rightarrow CO_2(g)$$

2 More coke reduces carbon dioxide making carbon monoxide.

$$C(s) + CO_2(g) \rightarrow 2CO(g)$$

3 Carbon monoxide moving up the furnace reduces iron ore falling down the furnace.

$$Fe_2O_3(s) + 3CO(g) \rightarrow 2Fe(l) + 3CO_2(g)$$

4 Unreacted iron ore is reduced by unreacted coke.

$$2Fe_2O_3(s) + 3C(s) \rightarrow 4Fe(l) + 3CO_2(g)$$

5 Limestone decomposes $CaCO_3(s) \rightarrow CaO(s) + CO_2(g)$

6 Calcium oxide reacts with high melting point impurities forming slag, which also protects the molten iron from the air blast.

$$CaO(s) + SiO_2(s) \rightarrow CaSiO_3$$

Assessing pupils' learning

Ask pupils to write a simplified technical guide to the blast furnace.

Rusting

Rust is a complex mixture of substances, but most of the iron in rust is in the form of hydrated iron(III) oxide. The reaction proceeds in a moist atmosphere with the iron(II) ions reacting to form iron(III) ions. The reaction is accelerated by atmospheric sulfur dioxide or sulfuric acid in urban areas, and by Cl^- in coastal areas. It is prevented by plating, painting or alloying. Rusting is negligible in hot, dry areas as the critical humidity is too low.

Rusting is an example of a chemical reaction which could form the basis of an investigation at KS3. In order to obtain some meaningful qualitative results you could use rust indicator which will give a colour change when rusting takes place. Rust indicator is made up of approximately one spatula of potassium hexacyanoferrate in 50 cm^3 of very dilute sodium chloride solution. It can be stored in a dropping bottle. The solution will change from yellow/green to dark blue/green when Fe^{2+} ions are released during rusting.

Alloys

Metallurgists formulate mixtures of metals called **alloys**. These alloys have different properties from the constituent metals and therefore increase the range of materials available to the metal technologist. Pupils being entered for higher GCSE papers will need to make the link between the structure and properties of metals and alloys. The alloying of metals usually involves adding a few atoms which are of a different atomic radius. The larger atoms prevent the layers from sliding over one another and so the alloy is harder than the original metals (see Chapter 3).

Table 9.5 Common alloys

Brass	Zn	Cu
Bronze	Sn	Cu
Pewter	Sn	Cu
Steel	Fe	C
Stainless steel	Fe	Cr, Ni, C
Cupro–nickel	Cu	Ni
Duralumin	Al	Cu, Mg

Different sizes of polystyrene balls or marbles can be used to represent the crystalline structure of metals in rows. Nuffield *Pupils' Books* and Nuffield *Co-ordinated Chemistry* have excellent coloured pictures of metal structures. The molecular modelling section of the *Chemistry Set* CD-ROM could be used to illustrate this.

An effective demonstration of alloying in the laboratory uses copper coins.

S **Safety Advice:** This is a teacher demonstration only. You should wear eye protection and use a safety screen.

- Heat 24 g of sodium hydroxide and 5 g of zinc powder in 100 cm^3 of water to boiling point.
- Place the copper coin carefully on top of the zinc in the beaker.
- Leave for a few minutes until the coin appears silver.

- Remove the coin which is now plated in zinc.
- Wash off excess zinc under a cold tap.
- Hold the plated coin in the hottest part of the Bunsen burner.
- The copper coating changes to brass as the copper atoms in the coin move to the surface and form an alloy with the zinc atoms coating the coin.

Analysing metal ions

Some chemistry syllabuses require that pupils are able to identify metal ions in solution using sodium hydroxide solution (see Figure 9.2). The SATIS unit 1203 *Prospecting with Chemistry* uses this method to solve a geological problem.

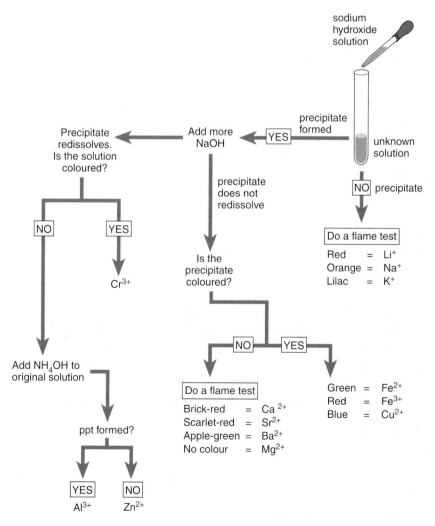

Figure 9.2 *Testing for metals*

ENRICHMENT AND EXTENSION ACTIVITIES

Information communication technology

Use a publishing package to produce a pamphlet/leaflet to explain to visitors what they can expect to see during a visit to a) a blast furnace or b) an aluminium smelting plant. ISCOR, http://www. ISCOR.com/vdb/production.htm

Produce a flow diagram comparing and contrasting the extraction of Al, Cu and Fe.

Use the *Chemistry Set* CD-ROM to search for data on metal properties and reactions.

Practical activities and investigations

Investigate the metal found in a drawing pin, staple and paper clip.

Which metal is best at protecting an oil rig against corrosion?

Which combination of metals produces the biggest voltage using fruit as an electrolyte?

Prepare two different samples of copper metal from copper compounds by different methods. (Pupils are provided with copper(II) oxide and copper(II) sulfate solution, carbon powder, access to lab equipment and Zn powder on request.) Suggestion: displacement of copper from $CuSO_4$ using Zn, and reduction of CuO using copper powder.

References

Teacher resources

Biblical Metals – SATIS 16–19 part 60, considers metal extraction and metal working, properties of metals and alloys, and uses of metals or alloys in the past and today.

The Earth's Resources: Metals – colour resource produced by RTZ which looks at mining, extraction and uses of common metals.

SATIS 1203 – *Prospecting by Chemistry*, is a simulated geochemical exploration for iron ore involving the chemical analysis of water samples. It involves reading material, experimental investigation, writing reports, data handling and interpretation.

Science Focus: The Salters' Approach – Metals. Spot-the-difference analysis of metals involving practical skills and problem solving.

Chemistry in Action – Rusting All Over the World – corrosion and the environment activity from CIEC York.

SATIS 310 – *Recycling Aluminium*

SATIS 103 – *Controlling Rust*

SATIS 604 – *Metals as Resources*

Nuffield Co-ordinated Science Teachers' Guide – analysing an alloy

Books

Lewis, M. *Revise through Diagrams.* Oxford University Press

Starting Science, Oxford University Press

chapter 10

Useful products from air and changes to the atmosphere

BACKGROUND

The Earth's atmosphere is a relatively thin layer of mobile, highly reactive gas, essential for life. This layer extends approximately 400 km above the surface. Our current atmosphere is an oxidising environment and is very different from the original atmosphere which did not contain much oxygen and was strongly reducing. The original atmosphere was probably similar in composition to solar nebula – rich in hydrogen and helium and close to the present atmosphere of the giant planets which contain large amounts of methane and nitrogen. The lighter elements of the primeval atmosphere were lost to space and replaced by compounds outgassed from the crust (or according to some more recent theories, from the impact of comets and other materials). Water vapour condensed forming oceans and lakes into which carbon dioxide dissolved and reacted to form carbonate rocks.

It is widely believed that lightning provided enough energy to produce amino acids which combined to make proteins. Bacteria are believed to have evolved in the primordial 'soup' in which primitive plants developed. The oxygen characteristic of our atmosphere was almost all produced by blue-green algae which grew in this soup around 2000 million years ago. Marine organisms subsequently developed in the oceans, absorbing carbon dioxide to form hard shells.

Oxygen reacts with other materials very readily, but the excess amounts produced by emerging biological systems created an oxygen-rich environment enabling aerobic life forms to evolve rapidly. Without free oxygen, life as we know it would not have evolved on the planet. At the same time, oxygen was altered in the high atmosphere to form a layer of ozone which protected the

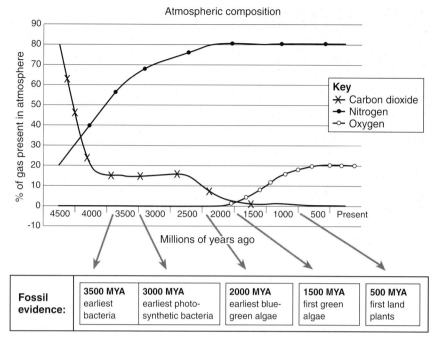

Figure 10.1 *Evolution of the atmosphere*

surface from damaging UV rays. As a result, higher order plants and animals were able to evolve on the surface of the Earth.

The present composition of the atmosphere is 79% nitrogen, 20% oxygen and 1% other gases.

The atmosphere can be divided into several distinct layers.

- The **troposphere** is the lowest level of the atmosphere and the prevailing weather conditions on the surface are set in this layer. It is a region of rising and falling packets of air and the highest clouds are found here.
- The **stratosphere** is above the troposphere and contains very little air which flows mostly in a horizontal direction. Aircraft travelling through the stratosphere seal in the air for passengers to breathe and have efficient heating systems because outside temperatures in this area drop to −550°C. The high level ozone layer is found in the upper reaches of the stratosphere.

Many pressure groups are concerned that the composition of the atmosphere is being altered as a direct result of Man's activities.

KEY STAGE 3 CONCEPTS

Some pupils do not consider gases to have material character and assign them transient properties. Others do not consider gas to have mass and view it as something that simply rises and floats. It is important that pupils develop the idea that gases have mass so that they will understand why mass is conserved in chemical changes that involve gases as reactants or products. Most 11- and 12-year-old pupils recognise that air is needed for burning but few recognise that it is a reacting substance.

> ### Assessing pupils' learning
>
> Ask pupils to produce an alien's guide to the Earth's atmosphere.
>
> Able pupils at KS4 could be asked to produce a report to outline the evolution of the atmosphere using the web pages suggested at the end of the chapter as a source of reference material.

Air has mass and takes up space

To try to help pupils see that air is a gas and is all around them you could set up a circus of activities which illustrates this. There are some suggestions in Chapter 1 which you could use and three further activities in Figure 10.2 which illustrate that gases take up space and have mass.

Composition of air and the extraction of gases from it

There are a number of activities which can be used to investigate air.

- You could show a video sequence to show how gases are extracted from air by fractional distillation or use the BOC Gas poster series.
- To show that air contains about 20% oxygen you could demonstrate the effect of passing air over hot copper turnings. Instructions for this can be found in most chemistry textbooks.
- To set the discovery of oxygen in context you could show pupils the BBC *Oxygen* video which considers the work of Lavoisier, Priestley and Scheele. This could be followed up by asking pupils to write letters from Priestley to Lavoisier.

Uses of major gases

Most chemistry books have information about the uses of gases in the air or you could search the *Chemistry Set* CD-ROM or the Internet.

(a)

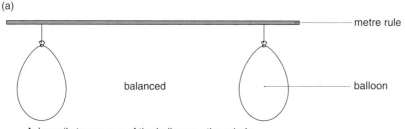

metre rule

balanced

balloon

Ask pupils to pop one of the balloons – the ruler's
equilibrium will change showing that air has mass.

(b)

gas takes up space

Ask pupils to suggest a method which would allow the
liquid to pour into the conical flask.

(c)

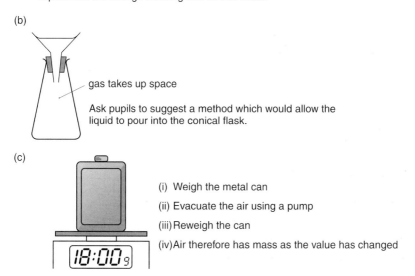

(i) Weigh the metal can

(ii) Evacuate the air using a pump

(iii) Reweigh the can

(iv) Air therefore has mass as the value has changed

18:00 g

Figure 10.2

There are a number of active ways of consolidating this information:

- produce a mobile to hang in the class showing the uses
- devise a game about gases along the lines of Trivial Pursuits™
- produce a wall display
- produce an information pamphlet
- devise a gas quiz, divide the class into small groups and ask them to produce
  questions to test each other
- write a poem about the gas
- write a story about day in the life of the gas
- create a web page
- review a video like *Air* from the Open University
- communicate with other pupils through *Science Across the World*
- write a public announcement of the discovery of the gas from the discoverer's
  point of view.

Physical and chemical properties of gases

Oxygen is a reactive gas. Most textbooks have instructions for class practicals involving heating substances to produce oxides. Rusting is one example of a reaction involving the production of an oxide – in this case, iron oxide. For more information on this reaction see 'Metals'.

Carbon dioxide is produced when sugars are broken down by yeasts in fermentation reactions or when bicarbonates are decomposed in a hot oven in cake mixes. Carbonated drinks also contain carbon dioxide. A fun demonstration which illustrates a property of carbon dioxide involves making a foam fire extinguisher. It is very messy so make sure you cover the bench beforehand.

S **Safety Advice:** Avoid getting foam in the eye. In addition, the foam will make the floor slippery so ensure that pupils direct their foam jets at a sink.

Making the foam fire extinguisher

- Put two spatula measures of aluminium sulfate into a beaker containing approximately 30 cm^3 of water. Stir vigorously.
- Add two spatula measures of sodium hydrogencarbonate to another beaker and add 30 cm^3 of water to this. Stir vigorously.
- Pour the aluminium sulfate solution into a conical flask and add a few drops of washing-up liquid.
- Carefully add the contents of the other beaker to the conical flask. Put a close-fitting bung and delivery tube drawn out into a fine nozzle into the conical flask containing the two reagents quickly and stand back!

> ### Assessing pupils' learning
>
> Ask pupils to use their knowledge and skills to investigate which gas is given off when Alka–Seltzer™ tablets are dissolved in water.

Changes to the atmosphere

There has been quite a lot written about global changes to the atmosphere. Pupils at Key Stage 3 should be introduced to the idea that the burning of fuels produces carbon dioxide and that some also produce sulfur dioxide from the sulfur impurities in the fuel. You can easily demonstrate this by igniting a match and holding a piece of damp litmus in the gases produced. The litmus will turn red as acidic sulfur dioxide is produced.

Nitrogen reacts with oxygen at the high temperatures found in car engines, particularly when they are ticking over in traffic jams. The nitrogen dioxide contributes to the problem of acid rain.

Table 10.1 Preparing gases in the laboratory

Gas	Reagents required	Apparatus required	Test for gas
Carbon dioxide	2 M hydrochloric acid added slowly to excess marble chips		Turns limewater cloudy
Oxygen	20 vol hydrogen peroxide added carefully to solid manganese(IV) oxide	Dry gas, CaCl$_2$	Relights a glowing splint
Hydrogen	1 M sulfuric acid added to zinc turnings or magnesium ribbon	magnesium ribbon, lighted splint	Makes a squeaky pop with a lighted splint

Investigating acid rain

At Key Stage 3 you could show how acidic gases affect metals or plant growth by setting up a practical which involves monitoring the growth of cress seeds or the effect on different metals in polluted environments. Set up two beakers with cress seeds growing on damp cotton wool or leave pieces of metal in the beaker. You can produce sulfur dioxide in the beaker by mixing a very small amount of sodium metabisulfite and citric acid in a small volume of water.

S Safety Advice: Gas preparation should be set up by the teacher. Sulfur dioxide is toxic by inhalation. Pupils with known breathing difficulties must not inhale the gas.

You could discuss with pupils the ways in which sulfur dioxide and nitrogen dioxide levels are reduced by using catalytic converters. Pupils could make a model of the converter using junk materials showing carbon dioxide, carbon monoxide and nitrogen dioxides coming out of the engine into the catalytic converter, and carbon dioxide, nitrogen gas and water coming out of the exhaust.

KEY STAGE 4 CONCEPTS

Fertiliser manufacture

You could co-ordinate this section with the biology course which also includes reference to the nitrogen cycle and eutrophication. Pupils could make a fertiliser, ammonium sulfate, in the laboratory using ammonia and sulfuric acid. The free publication *Investigating Water*, available from BNFL, gives detailed background information on water treatment and pollution including testing for nitrates.

The Haber Process

This reaction is examined in higher level papers. *Captains of Industry*, a resource pack for the 14–16 age group, is available from the Chemical Information Education Centre office and provides detailed background information on the economics and reactions set in an industrial context. The pack sets the reaction in a real context and may motivate more able students who are sometimes not engaged by practical work.

To investigate the Haber Process as an example of an equilibrium reaction you could use the simulation of the process on the *Chemistry School* CD-ROM available from New Media, or set up a spreadsheet to calculate yields of ammonia.

Local atmospheric pollutants

- **Acid rain** is caused by the reaction of sulfur dioxide and nitrogen dioxide in the atmosphere to form sulfuric acid and nitric acid. A lot of information, including a role play activity, is available in the *Acid Rain* units from the *Science Across the World* (SAW) and SATIS units. The SAW schools in Scandinavia are very willing to air their views on acid rain and you could subscribe to the project and write, fax or e-mail schools with your views. Pupils could make a video to send to the pupils in the other schools or produce a school newsletter.
- **Low level ozone.** Pupils are often confused by the difference between the high-level protective layer of ozone gas found in the stratosphere and the toxic low-level ozone found in urban areas. Many of the environmental pressure groups and the Department of the Environment produce glossy information pamphlets that could be used to supplement textbooks.

Global environmental pollutants

- **Ozone depletion** in the upper stratosphere. A lot has been written about this subject and pupils are often quite well informed. You could ask them to write an article about the problem or produce a page of frequently-asked questions with answers.
- Increased carbon dioxide levels – the **greenhouse effect**. This is an example of an area of science where the issues are not clear cut. There are differing views about the extent of the greenhouse effect and its ultimate affect on the climate. You could use this context to consider the idea of scientific certainty. Having a good source of information is vital if pupils are going to be able to decide on what is a scientific certainty and what is uncertain. A good starting point is the *Climate Change* pamphlet available free from the Natural Environment Research Council (NERC). There are many web pages which could be used and there is also a *Science Across the World* greenhouse effect unit. You could allocate 'expert' panel members from the class for a 'Question Time'.

You could also approach this area at a much less demanding level by asking pupils to audit their own carbon dioxide production. Tebbutt's *Spreadsheets in Science* book provides three templates at different skill levels. The program is available in both PC and Acorn formats and allows pupils access to the carbon dioxide simulation.

Assessing pupils' learning

Able pupils could read about the Gaia theory proposed by Dr James Lovelock, which suggests that Earth is in fact a self-regulating entity. Lovelock says that temperature, oxidation state, acidity and aspects of the rocks and waters are kept constant, and that this homeostasis is maintained by active feedback processes operated automatically and unconsciously by the biota. Pupils could research and report on this idea.

ENRICHMENT AND EXTENSION ACTIVITIES

Information communication technology

Understanding Reactions – the Haber Process, CD-ROM from New Media
Warwick Spreadsheet Package
E-mail schools through the *Science Across the World* programme.
Use a publishing package to produce the *Aliens Guide to the Atmosphere*.

Internet references

NERC Centre for Global Atmospheric Modelling:
http://www.met.rdg.ac.uk/ugamp/ugamp.html

UK Global Environment Research Office:
http://www.nerc.ac.uk/ukgeroff/welcome.htm

Hadley Centre for Climate Predication and Research:
http://www.meto.gov.uk/sec5/sec5pg1.html

Global warming: http://www.chem.ucl.edu/instruction/applets/canonical.html

Ozone depletion: http://uarsfot08.gsfc.nasa.gov/

Gaia: http://magna.com.au/~prfbrown/gaia_jim.html

Science Across the World: http://www.bp.com/saw

References

Video
ST240, *Air*, The Open University

Teachers resources *BOC Gases* poster series and teacher's booklet published by BOC Ltd.

Cleaning Up Your Act – Chemistry in Action from CIEC. A unit which considers various alternatives to cut down on emissions from car exhausts.

Climate Change Scientific Certainties and Uncertainties, a pamphlet published by the Natural Environment Research Council.

Books
Tebbutt, M. (1995) *Spreadsheets in Science*

3 Patterns of Behaviour

chapter 11

The periodic table

BACKGROUND

The periodic table is the trademark of every chemist and Peter Atkins summarised just how important it is when he said:

'The periodic table is arguably the most important concept in chemistry, both in principle and practice. It is the everyday support for students, it suggests new avenues for research to professionals, and it provides a succinct organisation of the whole of chemistry. It is a remarkable demonstration of the fact that the chemical elements are not a random clutter of entities but instead display trends and lie together in families. An awareness of the periodic table is essential to anyone who wishes to disentangle the world and see how it is built up from the fundamental building blocks of chemistry, the chemical elements. Anyone who seeks to be familiar with a scientist's-eye view of the world must be aware of the general form of the periodic table, for it is a part of scientific culture.'

It took nearly a hundred years to arrive at the current long form of the periodic table drawn up by Dimitri Ivanovich Mendeleev in 1872. The story of those who contributed to the current form is told very well in the BBC *Periodic Table* video which also has the Tom Leyer periodic table song as an accompaniment. Mendeleev was a visionary chemist. He was scoffed at by his colleagues because he left gaps in his table to allow for elements which he suggested had yet to be discovered. One of these elements, which he called ekasilicon, was discovered a few years after his death and later named germanium.

Table 11.1 Progression through the Key Stages

KS3 Level 3–5	KS3 Level 5–8/ Foundation GCSE	Higher GCSE
• Grouping of elements into the periodic table	• Increasing atomic number • Trends in groups 1, 2, 7 • Noble gases and transition metals	• Position in the table and how it is related to the outer electrons • Make predictions about elements based on their position in the periodic table

KEY STAGE 3 CONCEPTS

Grouping elements into a periodic table

At Key Stage 3 you could introduce pupils to the basic geography of the table. Using the *Chemistry Set* CD-ROM or one of the periodic table web pages ask pupils to annotate a copy of the table as suggested below.

Pupils could annotate the table to show

- metals and non-metals
- liquid elements
- solid elements
- gaseous elements
- the date of discovery of the element
- elements found in the body
- the most abundant elements.

KEY STAGE 4 CONCEPTS

Patterns in the periodic table

At Key Stage 4 pupils should go on to look at trends and patterns in the table. There is plenty of scope for practical work to illustrate the chemical properties and trends of each group. *Pathways through Science: Chemicals* has a series of practical activities which pupils can carry out safely. Most GCSE textbooks also offer suggestions. For those pupils who do not enjoy practical work, there are a number of trends to be 'discovered' by literature searches instead. You need to issue careful guidelines to ensure that their reading and written work is focussed.

Able pupils may find Peter Atkins' *The Periodic Kingdom* interesting. In it he treats the periodic table like a country. The book travels around the 'regions' introducing the basic chemistry of the table.

The key trends required by GCSE examinations are:

Pattern 1: metal/non-metal trends

You could point out to pupils that zinc and aluminium have **amphoteric** oxides – i.e. they can behave like acids in alkaline conditions and like bases in acidic conditions. Baby nappy-rash cream often contains zinc oxide for this purpose.

Pattern 2: down the group

The similar properties within a group are related to the outer shell electrons. Trends and differences going down a group are related to the number of inner electrons and hence the size of the atoms. The atoms of the elements at the top of a group are often so small that they have unusual properties.

Pattern 3: across the periods

Take a long strip of paper and divide it into 109 equal sections. In each section write the names of the elements in turn. Label each one with its atomic number. Use colour to indicate elements which have similar chemical and physical properties as given in Table 11.2.

Table 11.2 Properties of metals and non-metals

	Physical	Chemical
Metals	Conduct	Basic oxides
	Ductile/malleable	Reducing agents
		Form cations
Non-metals	Insulate	Acidic oxides
	Brittle	Oxidising agents
		Form anions

Assessing pupils' learning

Ask pupils to write an article summarising the key trends of the periodic table, or write a review of *The Periodic Kingdom*.

Electronic structure and the periodic table/higher level GCSE

Ask pupils to use the periodic table to write the electronic structure of the first 20 elements onto a blank grid, then look for patterns. The main point to make here is that elements in the same group of the periodic table have the same number of electrons in their outer shell. Explain that chemical properties are linked to outer shell electrons. Pupils could write the ions formed onto the blank grid in a different coloured ink.

The key ideas that pupils should know for higher level GCSE examinations are summarised in Table 11.3.

To help illustrate the point that reactivity increases down the group on the left-hand side of the table and up the group on the right-hand side of the table, you could carry out a role play activity. Ask two pupils to stand at the front of the lab. Label one the nucleus of lithium and the other close by as the outer shell electron of lithium. Using two other pupils, place one pupil adjacent to the lithium nucleus and label them caesium nucleus and place the other at the other far end of the room labelled caesium outer shell electron. Discuss the force of attraction between the electrons and their corresponding nuclei and try to link this with reactivity. Ask pupils to organise themselves into fluorine and iodine electron shells and link this to the opposite trend on the right-hand side of the table.

EXTENSION AND ENRICHMENT ACTIVITIES

Information communication technology

CD-ROM
Chemistry Set from New Media.
Chemistry Set 2000 from New Media.
Warwick Spreadsheet Package

Internet references
http://www.shef.ac.uk/~chem/web-elements/html. An excellent site.
http://www.chemsoc.org/viselements. An unusual site linking art, science and history of science.
http://www.bigfoot.com/~warwick-bailey
http://chemlab.pc. maricopa.edu/labbooks. An excellent site with downloadable obscure periodic tables.

References

Teacher resources
Pathways through Science: Chemicals, Longman, London

Chemistry Set: the Lessons, New Media

Table 11.3 Key trends of reactions in groups

Group 1 Alkali metals	Group 2 Alkaline earth metals	Group 7 Halogens	Group 0 Noble gases
• soft silvery metals • very reactive • form ions with a single positive charge Li^+ Na^+ K^+	• soft but not as soft as group 1 • reactive but not as reactive as group 1 • form ions with a double positive charge Mg^{2+} Ca^{2+}	• coloured • diatomic molecule • very reactive • form ions with a single negative charge F^- yellow gas Cl^- green gas Br^- red liquid I^- grey solid	• colourless gases • found in the air in small amounts • very unreactive

Going down the group:

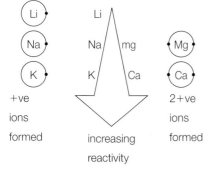

• melting point decreases • reactivity increases	• reactivity increases	• reactivity decreases; each halogen displaces the one below it from their compounds	• elements get more dense • boiling points increase

+ve ions formed

increasing reactivity

2+ve ions formed

increasing reactivity

He
Ne
Ar
Kr
Xe

increasing density

outer electrons getting further from nucleus

smaller atoms at the top of the table attract electrons more strongly and so are more reactive

Reactions:

• react vigorously with water • compounds soluble	• react with water but less rapidly	• react with hydrogen • react with metals

Books

Atkins, P. (1995) *The Periodic Kingdom.* Weidenfeld and Nicolson, London

Elmsley, J. (1987) *The Elements.* Oxford University Press, Oxford

Levi, P. (1987) *Periodic Table.* Abacus Books

chapter 12

Acids and bases

BACKGROUND

The word 'acid' is derived form the Latin word *acidus* meaning sour. An acid, at a very simple level, is capable of forming hydrogen ions when dissolved in water. These aqueous solutions have the following properties:

- a sharp taste
- turn litmus red
- liberate carbon dioxide from a metallic carbonate
- react with zinc or similar metals to produce hydrogen gas
- neutralise bases.

Attempts at defining an acid at a fundamental level began in the 18th century with Lavoisier, although his ideas did not prove to be accurate and it is the work of Svante August Arrhenius in 1887 which is recognised as being the most influential. Arrhenius proposed that an acid is any substance that produces hydrogen ions, H^+, in water and that a base is anything that produces hydroxide ions, OH^-, in water. For most pupils at GCSE this is a suitable model to use to explain the chemical properties. Beyond GCSE, pupils will encounter other explanations, such as the Brönsted–Lowry theory developed in 1923. The Brönsted–Lowry theory suggested that an acid is any substance that can transfer a proton to another substance. A base is defined as any substance that can accept a proton.

The term 'acid' was extended by Gilbert Lewis in the 1930s to include substances which are electron acceptors. For example, aluminium chloride can accept a pair of electrons from a chloride ion forming the $AlCl^{4-}$ ion, which is a **Lewis acid**.

The term 'pH' was developed in 1909 by the Danish chemist Sorensen who worked in the Carlsberg brewery in Copenhagen. He proposed that concentrations of H^+ be treated as an exponential value. He abbreviated the hydrogen ion concentration, for example 10^{-1} became pH 1, and in doing so arrived at the pH scale 1–14 now in common use. The letter 'p' comes from a word which was common in Germanic and Roman languages, *potenz* meaning power.

Table 12.1 Progression through the Key Stages

KS3 Level 3–5	KS3 Level 5–8/ Foundation GCSE	Higher GCSE
• pH scale	• Reactions of acids	• Hydrogen ion concentration
• Use of indicators	• Making salts	• Strength and concentration
	• Neutralisation	• Reactions of concentrated acids
	• Acid rain	

KEY STAGE 3 CONCEPTS

Pupils' ideas of acids at KS3 are derived from

- sensory experiences, for example lemons taste sour
- crime stories about acid baths and corpses
- anti-acid remedies
- acid rain.

Pupils do not often connect the reactions of acids with chemical properties and are often confused by the different models of acids and bases offered to them as they progress through school. Although the words and ideas change their meaning from one model to another, pupils are often not made aware of this because the teacher does not introduce the new model as precisely that.

Bases are less well known in daily life and pupils are often able to name acids but not bases.

Indicators and neutralisation at KS3

There are still schemes of work in use at KS2 which refer to the previous National Curriculum and primary science courses often teach pupils about indicators. You may find that pupils in your class are already aware of how plant extracts can be used to test for acids and have used litmus. You may like to extend the planned work for these pupils or start at a different point in your scheme of work. Equally, some pupils may not have used indicators. Figure 12.1 offers a teaching strategy which would allow pupils to decide their own starting point.

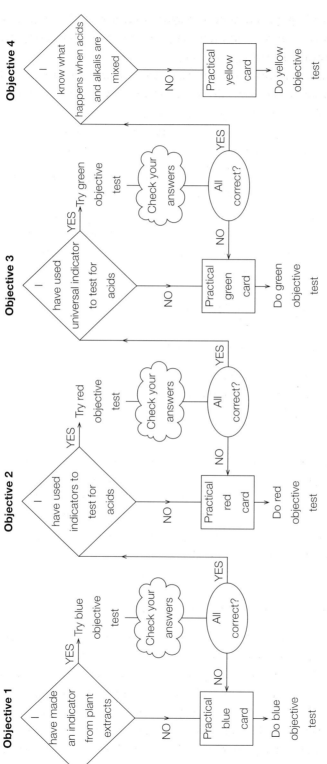

Figure 12.1 *Progression of activities. Differentiation for pupils from different primary schools*

Frequently, pupils are quite autonomous learners in primary schools and are sometimes held back by the whole class teaching approach found in some science classrooms at KS3.

Secondary science teachers often cite safety as the main reason why they must direct classroom practical activities very closely. With close supervision and careful induction this need not be an issue.

A suggested progression of activities for use with the teaching strategy suggested in Figure 12.1 could be:

1 using red cabbage or coloured flowers as an indicator, for example hydrangea flowers

2 using commercially-produced plant indicators such as litmus

3 familiarisation of the colour change of common narrow range indicators such as litmus and phenolphthalein

4 introduction to full range universal indicator

5 familiarisation with the pH scale

6 pH of household substances

7 demonstration of a neutralisation reaction in a burette (Figure 12.2)

Assessing pupils' learning

- Starting from universal indicator solution, water, 0.1 mol dm^{-3} hydrochloric acid solution and 0.1 mol dm^{-3} sodium hydroxide solution only, ask pupils to produce red, orange, yellow, green, blue and violet solutions and write a reliable recipe for creating the colours.
- Using a weak citric acid solution labelled 'bee sting' and a weak solution of sodium bicarbonate labelled 'wasp sting', ask pupils to investigate the best household substance which could be used as an antidote to each sting. Pupils could produce an information pamphlet which could be inserted into the first aid box.

Reactions of acids

Starting Science has a good section containing pupil activities which look at the reactions of acids although most KS3 textbooks suggest practical activities to illustrate this area.

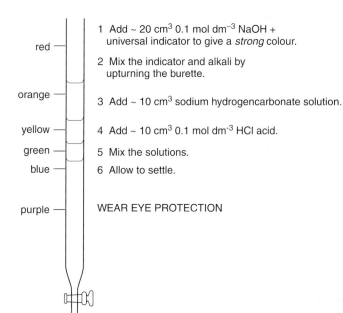

red —
1 Add ~ 20 cm³ 0.1 mol dm⁻³ NaOH + universal indicator to give a *strong* colour.

2 Mix the indicator and alkali by upturning the burette.

orange —
3 Add ~ 10 cm³ sodium hydrogencarbonate solution.

yellow —
4 Add ~ 10 cm³ 0.1 mol dm⁻³ HCl acid.

green —
5 Mix the solutions.

blue —
6 Allow to settle.

purple —
WEAR EYE PROTECTION

Figure 12.2 *A neutralisation demonstration*

KEY STAGE 4 CONCEPTS

Reactions of acids and making salts

The reactions of acids are summarised in Table 12.2. Most GCSE textbooks and schemes of work will provide practical details for these reactions.

S **Safety Advice:** Do NOT react Li, Na or K with acids.

Table 12.2 Reactions of acids

Alkali	acid + alkali → salt + water
	$HCl + NH_4OH \rightarrow NH_4Cl + H_2O$
Metal oxide	acid + metal oxide → salt + water
	$H_2SO_4 + CuO \rightarrow CuSO_4 + H_2O$
Metal hydroxide	acid + metal hydroxide → salt + water
	$2HNO_3 + Zn(OH)_2 \rightarrow Zn(NO_3)_2 + 2H_2O$
Metal carbonate	acid + metal carbonate → salt + water + carbon dioxide
	$H_2SO_4 + CaCO_3 \rightarrow CaSO_4 + H_2O + CO_2$
Metal	acid + metal → hydrogen gas + salt
	$2HCl + Mg \rightarrow H_2 + MgCl_2$

Salts are substances in which the hydrogen of an acid has been replaced by a metal. Each acid can produce a family of salts.

- Hydrochloric acid produces *chlorides.*
- Nitric acid produce *nitrates.*
- Sulfuric acid produces *sulfates.*
- Ethanoic acid produces *ethanoates.*

Assessing pupils' learning

Pupils could investigate which is the best method to produce copper sulfate. More able pupils could be asked to produce

- a moist white solid
- a blue solution
- a black powder

from the following starting materials:

- barium chloride solution
- dilute sulfuric acid
- copper carbonate powder.

and then explain the chemical reactions involved.

Strength and concentration

These concepts need only be introduced to pupils entered for higher GCSE exams.

The concentration of an acid is a measure of how much acid is dissolved in a known volume of liquid. The molarity (M) of a solution is the number of moles dissolved in 1000 cm^3 of water, so 2 M acid is twice as concentrated as 1 M acid. Strength is a measure of the extent to which the acid molecule

dissociates in water. If an acid is said to be weak it means that the acid molecule does not ionise completely in water. Ethanoic acid is an example of a weak acid.

The ethanoic acid molecule and ion reach dynamic equilibrium with the equilibrium normally well to the left, so there is little H$^+$ present.

In contrast, strong acids, like hydrochloric acid, dissociate completely and all the molecules are ionised.

$$HCl \rightarrow H^+ + Cl^-$$

The pH of the acid is a measure of strength. Strong acids have pH values of 1 while weak acids have pH values of around 3.

Figure 12.3 suggests a demonstration to illustrate this concept. The red colour of the solution becomes paler as it is diluted but the red colour does not actually change to orange and so the pH is not altering.

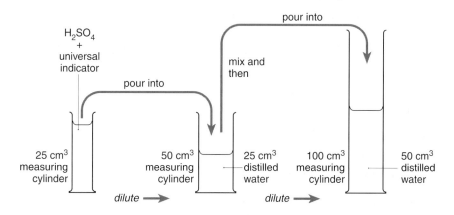

Figure 12.3

ENRICHMENT AND EXTENSION ACTIVITIES

Information communication technology

Internet references

http://www.chem4kids.com/
http://198.110.10.57/Chem/Chem1Docs/WaterWineBeer.html
http://science.demon.co.uk/handbook

References

Teacher resources

SATIS 709 *Which anti acid is best*

SATIS 505 *Making salts*

Books

Jarvis, A. (1993) *Particle Patterns.* Oxford University Press, Oxford

Starting Science. Oxford University Press, Oxford

Examination Questions

1 SOLIDS, LIQUIDS, GASES AND THE PARTICLE THEORY

KS3

1 Fuel A is stored in tanks. It is not stored under pressure. It flows along a pipe.
Fuel B is stored under pressure in small cylinders. It is used by campers.
Fuel C is stored in sacks.

Is Fuel A a solid, liquid or gas?

Liquid

Does Fuel B come out of the cylinder as a solid, liquid or gas.

Gas

Give an example of each type of fuel

A is petrol or oil

B is natural gas

C is coal, coke or wood.

(QCA)

KS4

2 a) The diagram shows a burning candle. Which state is the candle wax at in the diagram at points A, B and C?

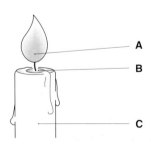

A *gas/vapour*

B *liquid*

C *solid*

b) i) The first box below shows the particles in a solid metal. Complete the other boxes to show the particles when the metal is a liquid and when it is a gas.

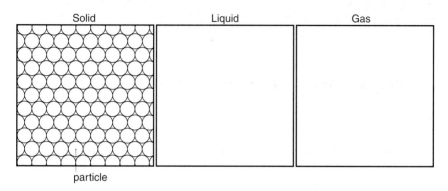

ii) When the solid metal is heated, it expands. Explain what happens to the particles as the solid metal expands.

· *particles gain energy*

· *vibrate more*

· *move apart.*

129

3 a) The diagram shows part of the water cycle.

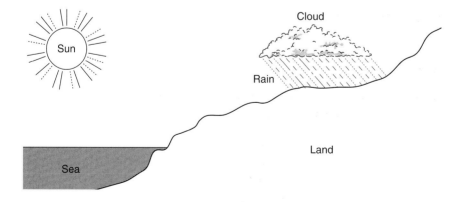

Explain the water cycle. In your answer write about:

how the sun starts the water cycle;

how the clouds and rain are formed;

how the rainwater goes back to the sea.

· *sun heats/evaporates the water/sea*

· *as water (vapour) rises it cools*

· *water vapour condenses*

· *large 'droplets' fall/precipitate as rain*

· *rainwater returns to sea by soaking through land/rivers*

2 ELEMENTS, COMPOUNDS AND MIXTURES

KS3

1 Air contains nitrogen, oxygen, argon, some water vapour and a little carbon dioxide. Complete the table for each substance

Substance	It is an element	It is a compound	It is a mixture	Number of atoms in one molecule
Nitrogen	✓			*2 (N_2)*
Oxygen	✓			*2 (O_2)*
Water vapour		✓		*3 (H_2O)*
Carbon dioxide		✓		*3 (CO_2)*

(QCA)

KS4

Foundation

2 Petrol is burned in air in car engines.

Gases in the air:
nitrogen
oxygen
carbon dioxide
argon
water vapour

Gases leaving the exhaust:
nitrogen
oxides of nitrogen
oxides of carbon
argon
water vapour

a) Put the gases in the air under the correct heading in the table.

Element	Compound
N_2	CO_2
O_2	H_2O
Ar	NO_x

b) Petrol is a mixture of hydrocarbons. Explain the word *mixture*.

compounds and/or elements together but not joined/bonded

c) Which gas in the air reacts with petrol in the engine?

oxygen

Higher

3 Rock salt contains insoluble solids and the soluble salt, sodium chloride. Arrange the following processes needed to separate sodium chloride from rock salt, in the correct order.

addition of water   crystallisation   evaporation   filtration   stirring

Explain the purpose of each process.

First process   *addition of water*
Purpose   *to dissolve soluble salt/sodium chloride*

Second process   *stirring*
Purpose   *to speed up formation of solution*

Third process   *filtration*
Purpose   *to remove insoluble solids*

Fourth process   *evaporation*
Purpose   *to remove some of the water/form a supersaturated solution*

Fifth process   *crystallisation*
Purpose   *separate sodium chloride/soluble salt*

131

3 ATOMIC STRUCTURE AND BONDING

KS3

1 Naturally-occurring atoms of the element sodium have atomic number 11 and mass number 23. Using this information, complete the table opposite.

Type of particle	Number in naturally-occurring sodium
Proton	11
Neutron	12
Electron	11

b) Another isotope of sodium has the mass number 24.

i) How is an atom of this type of sodium different from naturally-occurring sodium?

It has an extra neutron or has 13 neutrons.

ii) Despite the different mass number, these atoms are still sodium atoms. Explain why.

They have same number of protons or there are still 11 protons.

OR   They have the same number of electrons, there are still 11 electrons

and they have the same atomic numbers.

(QCA)

KS4

Higher

2 a) Use your periodic table to help identify the atoms A and B presented in the diagram below.

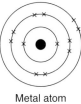

Metal atom
A

Non-metal atom
B

b) What happens to these electron arrangements when magnesium reacts with chlorine to form magnesium chloride, $MgCl_2$?

Mg atom loses two electrons/forms 2,8

Cl atoms gain one electron/forms 2,8,8

c) The compound magnesium chloride has *ionic bonding*. Explain what this means.

Mg ion positively charged/2^+ and Cl ion negatively charged/1^-.

Oppositely charged ions attract

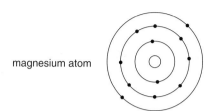

3 The diagram represents the arrangement of electrons in a magnesium atom.

a) Complete the table.

magnesium atom

	Number of			**Electron arrangement**
	protons	**neutrons**	**electrons**	
Magnesium-24	*12*	*12*	*12*	*2,8,2*
Oxygen-16	*8*	8	8	*2,6*

b) Magnesium oxide contains ionic bonding. Explain fully in terms of transfer of electrons and the formation of ions, the changes which occur when magnesium oxide is formed from magnesium and oxygen atoms.

Mg loses 2 electrons to form Mg^{2+}

O gains 2 electrons to form O^{2-}

c) Sodium chloride and magnesium oxide have similar crystal structure and both contain ionic bonding. The melting points of sodium chloride and magnesium oxide are 800°C and 2800°C, respectively. Suggest why the melting point of magnesium oxide is much higher than the melting point of sodium chloride. (Sodium chloride contains Na^+ and Cl^- ions.)

Attractive forces between Mg^{2+} and O^{2-} are much greater than between Na^+ and Cl^-.

More energy is needed to overcome attractive forces.

(MEG/OCR)

4 GEOLOGICAL CHANGES

KS3

1 This key can be used to identify some common crystalline rocks.

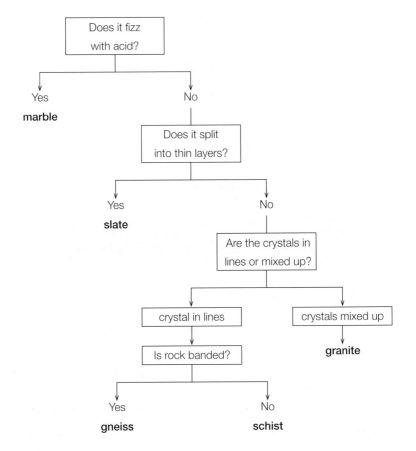

a) Use the key to describe *three* features of the rock called gneiss.

does not fizz with acid, does not split into thin layers, rock banded,

crystals in lines

b) Which other rock shown in this key is *most* like gneiss?

Schist

2 The diagram shows a cross-section of a type of volcano.

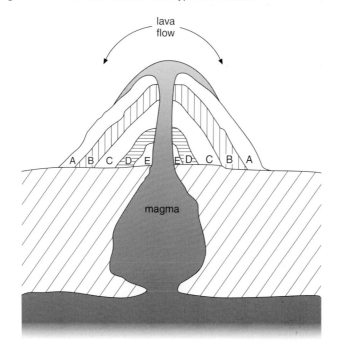

lava
flow

A B C D E E D C B A

magma

a) What type of rocks are produced by volcanoes?

igneous

b) Which of the labelled layers, **A**, **B**, **C**, **D**, or **E**, is the oldest?

Layer *E*

c) Explain why the crystals on the surface of rock layer **A** are smaller than the crystals in the middle of the same layer.

Colder on the surface, smaller crystals formed on rapid cooling, or rate of
cooling slower in the middle of the layer and hence crystals are larger

(MEG/OCR)

KS4

Foundation

3 a) What type of rocks are formed by the deposition, burial and compression of weathered rock fragments?

sedimentary

b) Below is a flow diagram of the rock cycle. Label the empty boxes using the words below.

igneous   metamorphic   sedimentary

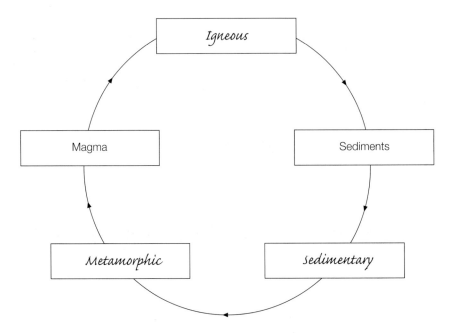

c) Limestone and marble are both forms of calcium carbonate. Limestone may contain fossils. Marble never contains fossils. Explain why.

limestone formed marble at high temperatures/pressure caused structural changes/destroyed fossils

Higher

4 Convection currents occur in the molten mantle of the Earth. The diagram opposite represents the structure of the inside of the Earth.

a) Describe what you might expect to find at **Y**.

new rock being formed; volcanoes

b) Suggest *two* effects that the convection currents would have on the ocean plates.

move apart; earthquakes at other edge

(MEG/OCR)

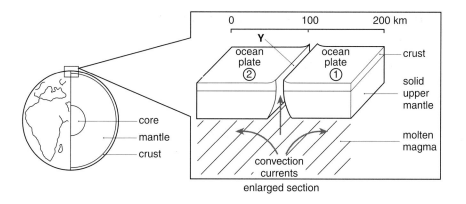

enlarged section

5 CHEMICAL REACTIONS

KS3

1 The diagram shows five test tubes, each containing exactly 12.4 g of copper(II) carbonate.

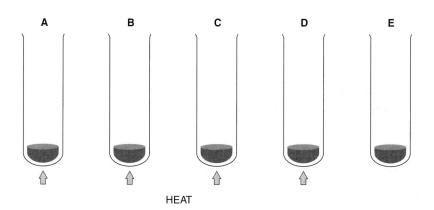

HEAT

When copper(II) carbonate is heated the following reaction occurs.

$$CuCO_3 \rightarrow CuO + CO_2$$

A pupil was given the following instructions

1 Heat each tube, cool it and then weigh it.
2 Repeat the process three times for each tube.
3 Calculate the mass of solid in each tube after each weighing.

The pupil forgot to heat one of the samples and made an error in weighing another of the samples. The results obtained are given in the table.

Sample	Mass of solid after heating/g			
	1	**2**	**3**	**4**
A	8.6	8.5	8.0	8.0
B	9.8	9.5	8.5	8.0
C	16.0	9.7	9.1	8.0
D	8.0	8.0	8.0	8.0
E	12.4	12.4	12.4	12.4

Use the results in the table to answer the following questions.

a) Which tube did the pupil forget to heat?

E

b) Which set of results contains the weighing error?

Sample C

Give a reason for your answer.

the mass has more than was started with

c) Why do you think there is no change in mass in results 3 and 4 in the case of Sample A?

the reaction had been completed

d) In which tube has most copper(II) carbonate decomposed after the first heating?

D

e) From the results how much copper(II) oxide can be obtained from 12.4 g of copper(II) carbonate?

8 g

(MEG/OCR)

KS4

Higher

2 Ammonia, NH_3, is made by reacting together hydrogen and nitrogen in the presence of iron. This reaction is called the Haber Process.

a) How does the presence of iron help the process?

catalyst

b) The table shows how much ammonia is produced using different conditions

	Percentage yield of ammonia at these temperatures		
Pressure/atm	**100°C**	**300°C**	**500°C**
25	91.7	27.4	2.9
50	94.5	39.5	5.6
100	96.7	52.5	10.6
200	98.4	66.7	18.3
400	99.4	79.7	31.9

From the values in the table, what happens to the yield of ammonia as:

i) the temperature is increased?

decreases

ii) the pressure is increased?

increases

c) Using ideas about particles colliding, explain how the rate of the reaction will change as the temperature increases.

moving faster, more collisions or 'harder' collisions, greater chance of successful collisions

d) The Haber Process is usually carried out at a higher temperature than the one which would give the highest yield. Suggest a reason for this.

goes faster

(MEG/OCR)

3 In the body, poisonous hydrogen peroxide, H_2O_2, is decomposed by catalase to give oxygen. Hydrogen peroxide can also be decomposed in the laboratory using manganese(IV) oxide.

a) Suggest the name of the other product of the reaction (besides oxygen).

water

b) Which one of the following would you use to test for oxygen? Tick the correct box.

glowing splint ☑

limewater ☐

lighted splint ☐

universal indicator ☐

c) What is the job of the manganese(IV) oxide?

catalyst

d) Which of the graphs below represents what happens to the mass of the manganese(IV) oxide as the reaction proceeds? Give a reason for your answer.

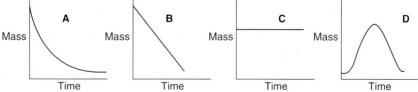

graph C because it takes no part in the reaction; it is not used up. It helps to decompose the hydrogen peroxide

e) Suggest why hydrogen peroxide decomposes faster with powdered manganese(IV) oxide rather than with a large lump.

powder has a greater surface area ....

(MEG/OCR)

6 ELECTROLYSIS

KS4

Foundation

1 Sodium chloride solution contains:

sodium ions, chloride ions, hydrogen ions, hydroxide ions

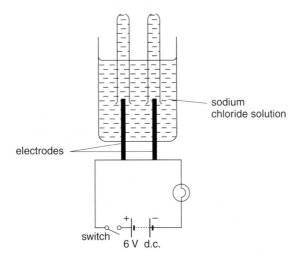

The diagram shows how sodium chloride solution can be split up using electricity.

a) Write down the name of this type of chemical reaction.

electrolysis ....

b) Describe what you would *see* at the electrodes when the electricity is switched on.

bubbles of gas at each electrode ....

gas collects in the tubes ....

c) Write down the names of the *two* gases made when the electricity is switched on.

hydrogen and chlorine ....

(MEG/OCR)

2 Sodium hydroxide (NaOH) is manufactured by the electrolysis of sodium chloride solution (NaCl). Hydrogen (H_2) and chlorine (Cl_2) are made at the same time. The equation below shows the overall reaction taking place.

$$2NaCl(aq) + 2H_2O(l) \rightarrow 2NaOH(aq) + H_2(g) + Cl_2(g)$$

The table below shows some of the uses of the products.

Product	Use
Sodium hydroxide	Making detergents, soap, paper, bleach and artificial fibres
Chlorine	Making bleach, solvents, pesticides and plastics, and to purify water supplies
Hydrogen	Making margarine and ammonia, and as a fuel

There are three main industrial processes for the electrolysis of sodium chloride solution. Information about these processes is shown in the table below.

Process	Information
Castner–Kellner	Produces a concentrated solution of sodium hydroxide but needs poisonous mercury that can contaminate water. The process also needs lots of energy.
Diaphragm	Produces a dilute solution of sodium hydroxide but the diaphragm needs replacing frequently. The diaphragm, which is made of asbestos, may cause cancer.
Membrane	Produces a concentrated solution of sodium hydroxide but needs pure sodium chloride solution to start with. The process needs very little energy.

Read the information carefully and use it to answer the questions.

a) What is the name of the raw material needed for all three processes which make sodium hydroxide?

sodium chloride/salt

b) The Castner–Kellner cell used to be the most common process to make sodium hydroxide. In recent years the membrane cell has replaced the Castner–Kellner cell in the manufacture of sodium hydroxide. Explain why.

Cell was energy expensive and involved the use of toxic mercury that contaminated water

c) In the membrane process chlorine (Cl_2) is formed at the positive electrode (anode).

i) Chlorine is formed from chloride ions (Cl^-). Write down the electrode reaction that occurs at the positive electrode.

$2Cl^-(aq) \rightarrow Cl_2(g) + 2e^-$

ii) Hydrogen (H_2) is formed at the negative electrode (cathode). The electrode reaction that occurs at the negative electrode is

$$2H^+ + 2e^- \rightarrow H_2$$

How many moles of hydrogen molecules (H_2) are produced when 1 mole of electrons (1 Faraday) is passed through sodium chloride solution?

0.5 moles Hydrogen (H_2)

iii) The relative atomic mass (A_r) of hydrogen is 1.0. What mass (in grams) of hydrogen is formed when 20 moles of electrons (20 Faradays) are passed through the solution?

20 moles of electrons produce 10 moles of hydrogen gas (H_2)

10 moles of hydrogen gas has a mass of 20 g

iv) Sodium chloride solution can also be electrolysed in the laboratory. Describe what you would *see* at the positive electrode.

bubbles of pale green gas evolved

d) Aluminium is manufactured by the electrolysis of *molten* aluminium oxide. Suggest *two* reasons why it is much cheaper to manufacture sodium hydroxide than to manufacture aluminium.

it is expensive in energy to melt aluminium oxide; a greater quantity of electricity is needed to produce 1 mole of aluminium than 1 mole of sodium hydroxide – (3^+ charge on Al ion)

(MEG/OCR)

3 a) A labelled diagram shown below is of a cell used to produce aluminium metal.

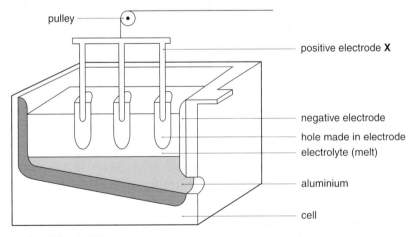

i) What is **X**?

anode/carbon

ii) Write a balanced ionic equation for the reaction taking place at the cathode.

$Al^{3+} + 3e^- \rightarrow Al$

iii) Name the gas that is also made in the cell.

oxygen

iv) **X** is hung on a pulley so it can be lowered. Why might **X** have to be lowered while aluminium is being produced?

level of melt goes down as aluminium is siphoned off

b) Pure copper is obtained by electrolysis using a thin, pure copper cathode and copper sulfate solution. A current of 200 amperes (A) is used for 12 hours. What mass of copper is formed at the cathode? Include the equation you are going to use. Show clearly how you obtain your answer and give the unit. (The Faraday constant (*F*) is 96 500 coulombs per mole (C/mol). The relative atomic mass of copper is 64.)

charge/coulombs = amps × seconds

coulombs = 12 × 60 × 60 × 200 = 8 640 000 C

copper ion has a charge of 2+

2 × 96 500 C gives 64 g of copper

8 640 000 C gives $\dfrac{64 \times 8\ 640\ 000}{2 \times 96\ 500}$

Answer 2865 g

7 EQUATIONS AND QUANTITATIVE CHEMISTRY

KS4

Higher

1 Zinc phosphide, formula Zn_3P_2, is found in some rat poisons. It is an ionic solid manufactured by heating zinc and phosphorus together.

a) i) Construct the chemical equation for the preparation of zinc phosphide.

$3Zn(g) + 2P(g) \rightarrow Zn_3P_2(g)$

ii) Predict the formula for the phosphide ion.

P^{3-}

b) Calculate the percentage by mass of phosphorus in zinc phosphide.

62/257 × 100

= 24.1%

c) In the factory, some of the zinc phosphide was accidentally contaminated with water. It forms zinc hydroxide and phosphine, PH_3. The equation is given below.

$Zn_3P_2(s) + 6H_2O(l) \rightarrow 3Zn(OH)_2(s) + 2PH_3(g)$

Calculate the minimum mass of water required to react completely with 257 kg of zinc phosphide.

1 mole zinc phosphate is 257 g and this reacts with 6 moles water which is

108 g so 257 kg of zinc phosphide would completely react with 108 kg of

water

d) A 3.10 g sample of phosphorus reacted with chlorine to form 20.85 g of a chloride of phosphorus. Calculate the empirical formula of the product.

$3.1\ g\ phosphorus\ \dfrac{3.1}{31} = 0.1\ mole$

the product must contain 20.85 $-$ 3.1 g = 17.75 g chlorine and this is

$\dfrac{17.75}{35.5} = 0.5\ moles\ chlorine$

simplest ratio is 0.1 P to 0.5 Cl which is PCl_5

(MEG/OCR)

2 A gaseous compound contains carbon and hydrogen only. It was found that 100 g of the compound contained 82.8 g of carbon.

a) i) Calculate the mass of hydrogen in 100 g of the compound.

17.2 g

ii) Use this information to calculate the number of moles of carbon atoms and of hydrogen atoms in 100 g of the compound.

carbon *82.8/12 = 6.9*

hydrogen *17.2/1 = 17.2*

iii) Use these figures to find the simplest formula for the compound.

b) The mass of 1 mole of the gaseous compound is 58 g. What is the molecular formula of the gaseous compound?

$C_2H_5 = (2 \times 12) + (5 \times 1)$

$= 29 \therefore formula = 29 \times 2$

$\therefore C_4H_{10}$

(MEG/OCR)

8 USEFUL PRODUCTS FROM OIL

KS3

1 a) Explain how crude oil was formed.

from 'sea creatures'/organic matter trapped

under high temperatures/pressures

over thousands/millions of years.

b) How is crude oil extracted from the ground?

by drilling

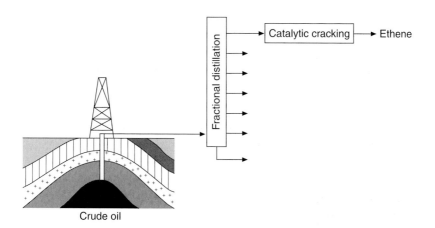

Crude oil

c) How can fractional distillation be used to separate compounds in crude oil?

heat crude oil

compounds boil at different temperatures

KS4

d) Outline how catalytic cracking can be used to produce ethene, C_2H_4.

thermal/heating

decomposition/larger molecules are broken into smaller molecules

145

KS3/KS4

Foundation

2 Crude oil is a mixture of different chemicals called hydrocarbons. Crude oil can be separated by heating it up and then collecting fractions at different temperatures. This apparatus can be used to do it.

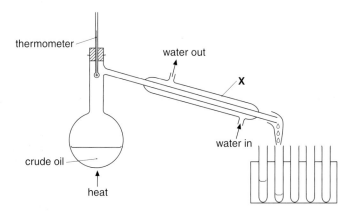

a) Write down the name of the method used to separate the crude oil.

fractional distillation

b) What is the job of part **X**?

to cool down the gases

condense into liquids

c) The table shows some of the properties of the fractions.

Fraction	Temperature in °C	How runny?	Colour	How it burns
A	up to 70	very	clear	easily, clean flame
B	70 to 150	fairly	pale yellow	fairly easily, a bit smoky
C	150 to 230	not very	yellow	difficult to light, a smoky flame

i) Another fraction was collected between 230°C and 300°C. What would it be like?

dark yellow/brown

not runny/thick

very difficult to light

very smoky flame

ii) Fraction **A** is used as a fuel in a car engine. Suggest reasons why Fraction **C** would be unsuitable for use in a car engine.

Hard to light; burns with a smoky flame; hard to vaporise; higher

viscosity

(MEG/OCR)

KS4

Foundation

3 a) The table below shows the formulae for some monomers and their corresponding polymers.

i) Complete the table.

ii) Name *one* monomer from the table which is a hydrocarbon.

ethene; propene

iii) What is the similarity between the structural formulae of the monomers?

all contain a double bond

Monomer	Structural formula	Polymer	Structural formula
Ethene	H H | | C=C | | H H	Poly(ethene)	$\begin{bmatrix} \text{H} & \text{H} \\ \| & \| \\ \text{C} - \text{C} \\ \| & \| \\ \text{H} & \text{H} \end{bmatrix}_n$
Chloroethene	H Cl | | C=C | | H H	Polychloro(ethene)	$\begin{bmatrix} \text{H} & \text{Cl} \\ \| & \| \\ \text{C} - \text{C} \\ \| & \| \\ \text{H} & \text{H} \end{bmatrix}_n$
Propene	CH_3 H | | C=C | | H H	Poly(propene)	$\begin{bmatrix} CH_3 & \text{H} \\ \| & \| \\ \text{C} - \text{C} \\ \| & \| \\ \text{H} & \text{H} \end{bmatrix}_n$
Tetrafluoroethene	F F | | C=C | | F F	Polytetrafluoro(ethene)	$\begin{bmatrix} \text{F} & \text{F} \\ \| & \| \\ \text{C} - \text{C} \\ \| & \| \\ \text{F} & \text{F} \end{bmatrix}_n$

iv) Write down the molecular formula of tetrafluoro(ethene).

C_2F_4

v) Poly(ethene) is used for many purposes instead of metals. Give *one* example where it is an advantage to use poly(ethene) instead of metal.

any reasonable response e.g. does not corrode, lighter

(MEG/OCR)

9 USEFUL PRODUCTS FROM METALS

KS3/KS4

Foundation

1 Iron ore is found in the Earth. This diagram shows how it can be made into iron metal.

a) Suggest *one* method which could be used to extract iron ore from the ground.

blasting, digging (open-cast) mining

b) Iron oxide is heated in a blast furnace to change it into iron metal.

i) Name a chemical heated in the blast furnace with the iron oxide.

coke, carbon or limestone

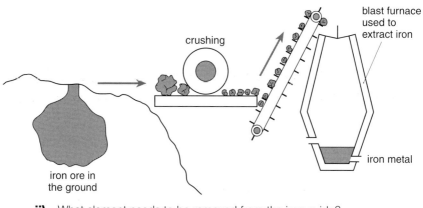

blast furnace used to extract iron

crushing

iron metal

iron ore in the ground

ii) What element needs to be removed from the iron oxide?

oxygen (oxide)

c) Underline *one* substance in this list which is a metallic ore.

bauxite charcoal diamond oil water

(MEG/OCR)

KS4

Higher

2 a) The diagram shows what happens when a small piece of potassium is placed on water which contains a small amount of a pH indicator.

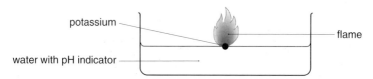

potassium

flame

water with pH indicator

i) How can you tell that potassium has a low density?

.......*floats on water* ....

ii) The reaction causes the pH indicator to change colour. What has caused this colour change? Explain your answer.

pH increases/pH is above 7

.......*alkaline solution formed because of potassium hydroxide/OH⁻* ....

iii) A flame is formed when potassium reacts with the water. Explain why.

.......*hydrogen released* ....

.......*ignited by exothermic/heat from reaction* ....

b) i) Helium, neon and argon are *unreactive* elements. Explain why.

.......*all have full outer shells of electrons* ....

ii) Draw a diagram to show the electron arrangement of an argon atom.

.......*2,8,8* ....

10 USEFUL PRODUCTS FROM AIR AND CHANGES TO THE ATMOSPHERE

KS4

Higher

1 The energy for some spacecraft rocket engines is produced by burning a mixture of liquid hydrogen and liquid oxygen. The burning fuel produces hot gases which are forced through a nozzle to produce lift. The diagrams show what happens.

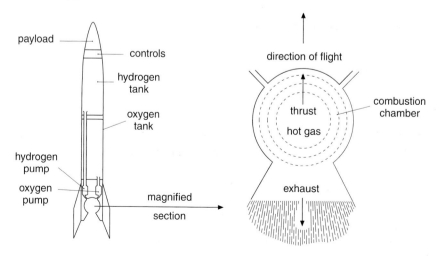

a) Give the equation for the reaction that takes place when hydrogen burns in oxygen.

Word equation   *hydrogen + oxygen → water(steam)*

Symbol equation   *$H_2 + O_2 \rightarrow H_2O$ or $2H_2 + O_2 \rightarrow 2H_2O$*

b) Using hydrogen and oxygen produces no pollution. Explain why.

Water is the only product and this is not a pollutant.

c) Which of the gases is the 'oxidiser'?

oxygen

d) The engines are less powerful when liquid air is used instead of liquid oxygen. Suggest a reason for this.

Concentration of O_2 in air is less; air has its oxygen diluted by other gases; air has less O_2 in it; air is a mixture of gases

e) Why does the total mass of the rocket decrease as soon as the fuel starts to burn?

the fuel is being used up

(MEG/OCR)

2 Ammonia is made by reacting nitrogen and hydrogen under the following conditions:

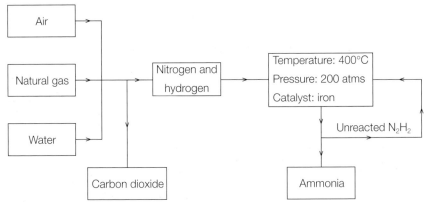

a) i)   Name the *two* sources of hydrogen in this diagram.

natural gas and water

ii)   Carbon dioxide can be removed by reacting the gas with an alkaline solution. Explain why.

forms a precipitate which can be filtered off

b) i)   Balance the equation below.

$$N_2 + 3H_2 \rightleftharpoons 2NH_3$$

ii)   What does ⇌ represent?

equilibrium reaction; the reaction can proceed in either direction

KS4

Foundation

1 a) In the periodic table the elements are arranged in an increasing order. What is it that increases?

atomic number

b) i) How does the reactivity of the elements change going down group 1?

more reactive

ii) Name the products of the reaction between any member of group 1 and water.

hydroxide and hydrogen

c) Look at group VII in the periodic table.

i) How many electrons are there in an atom of bromine?

Br 2,8,8,7

ii) How does the reactivity of the elements change going down group VII?

decreases

(MEG/OCR)

151

Higher

2 The abridged periodic table below shows the positions of the first 20 elements.

					H												He
Li	Be											B	C	N	O	F	Ne
Na	Mg											Al	Si	P	S	Cl	Ar
K	Ca																

a) In 1869, Dimitri Mendeléev published a table on which this periodic table has been based. He stated that 'when elements are arranged in order of atomic mass, similar properties recur at intervals'. Discuss this statement.

No – the periodic table is arranged in order of atomic (proton) number or some atoms of different elements have the same relative masses.

Yes – similar chemical properties do occur at intervals; these are the groups of the periodic table

b) i)   Why do the group 1 elements lithium, sodium and potassium have similar chemical properties?

have one electron in outer shell

ii)   The order of reactivity for the group 1 metals is:

potassium	most reactive
sodium	
lithium	least reactive

Using the periodic table, explain this order of reactivity.

reactivity increases in group with increasing atomic number/easiest to lose outer electron of potassium because of shielding effect of more inner shells.

12   ACIDS AND BASES

KS3/KS4

Foundation

1 a) i)   Which *one* of the following could be used to find the pH of a substance? Underline your answer.

litmus   limewater   starch   <u>universal indicator</u>

ii)   What problem would you have finding the pH of blackcurrant juice using this?

Coloured solution would affect indicator colour

b) The table shows the pH of solutions of five oxides, **V**, **W**, **X**, **Y** and **Z**. Find the table by writing in the description of the pH. Two have been done for you.

Oxide	pH	Description
V	1	Strong acid
W	6	*Weak acid*
X	7	*Neutral*
Y	9	Weak alkali
Z	13	*Strong alkali*

c) The treatment for stings usually involves neutralisation with a weak acid like lemon juice or a weak alkali like dilute ammonia. This stops the pain. The information below was found in a holiday guide.

> …if you step on a sea urchin in the sea, treat the sting with a mixture of olive oil and lemon juice…
>
> …a jellyfish sting should be treated with ammonia to stop the pain…

What does this information tell you about the pH of the stings of

i) sea urchins? *sting is alkaline*

ii) jellyfish? *sting is acidic*

(MEG/OCR)

KS4

Foundation

2 a) Dilute sulfuric acid has a pH of 2.

i) What is the difference between a dilute acid and a concentrated acid?
in the same volume/amount concentrated contains more acid/dilute contains more water

ii) Universal indicator solution is green when neutral. When added to dilute acid, the solution goes red. How could you make the solution go green again?
add an alkaline solution dropwise

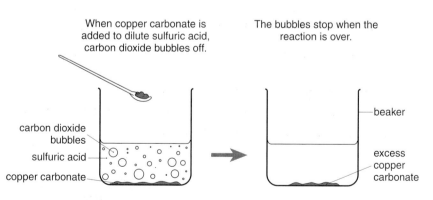

When copper carbonate is added to dilute sulfuric acid, carbon dioxide bubbles off.

The bubbles stop when the reaction is over.

beaker

carbon dioxide bubbles

sulfuric acid

copper carbonate

excess copper carbonate

b) i) Draw a diagram to show how copper carbonate can be removed when the reaction is over.

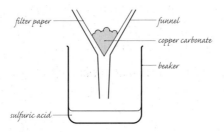

filter paper — funnel
— copper carbonate
— beaker
sulfuric acid —

ii) Name a simple test to show that the gas is carbon dioxide.

Test   *limewater*

Result of test   *goes white/milky*

iii) Complete the word equation for this reaction.

copper   + sulfuric ⟶ *copper* + *water* + *carbon*
carbonate   acid   *sulfate*   *dioxide*

iv) Name another copper compound that would react with dilute sulfuric acid.

copper oxide

Contact Addresses

TEACHERS RESOURCES

BP Educational Service, PO Box 934, Poole, Dorset BH17 7BR (01202) 669940 bpes@bp.com

Chemical Industry Education Centre (CIEC), Department of Chemistry, University of York, York YO1 5DD. (01904) 432523 ciec@york.ac.uk

Earth Science Teachers' Association: Geoff Nicholson, 28 Harthill Avenue, Leconfield, Beverley, E. Yorks. HU17 7LN

Institute of Petroleum, 61 New Cavendish Street, London W1M 8AR http://www.petroleum.co.uk/

Resources in Science Education, 18 Abercorn Road, Edinburgh EH8 7DJ

Shell Education Service, Business Services, PO Box 46, Newbury, Berks. RG14 2YX (01635) 31721

INFORMATION COMMUNICATION TECHNOLOGY

Chemistry Set, Chemistry Set 2000, Electrochemistry, Elements, Compounds and Mixtures, Understanding Reactions, Atom Viewer and *States of Matter.*
New Media Press Ltd, PO Box 4441, Henley-on-Thames, Oxon RG9 3YR. (01491) 413999; Fax: (01491) 574641 e-mail enquiries to new-media.co.uk http://www.new-media.co.uk

Warwick Spreadsheet System, Aberdare Publishing, 6 Nuthurst Grove, Bentley Heath, Solihull B93 8PD http://members.aol.com/aberdareco

Gas Equilibrium Modelling CD-ROM.
Newbyte Educational Software
(0141) 337 3355
http://www.newbyte.com/uk

INTERNET REFERENCES

http://www.chemists-net.demon.co.uk/software.html
Links to a variety of software such as periodic tables, clip art, molecular weight
calculators, a macro for MS Word to facilitate chemical equation writing and
other software of variable use.

http://dbhs.wvusd.k12.ca.us/Chem-History
Classic papers from the history of chemistry

BOC Ltd http://www.boc.com.gases

VIDEO SUPPLIERS

Channel 4 Schools, PO Box 100, Warwick, CV34 6TZ
eo@schools.channel4.co.uk
Online guides: http://www.channel4.com/schools/guides

Boulton–Hawker Films Ltd, Hadleigh, Ipswich IP7 5BG

BBC Videos for education and training
80 Wood Lane, London W12 0TT

OTHER CONTACTS

- The **Association for Science Education** (ASE) exists to improve the teaching of
  science. It is the largest of the subject teachers' associations.
  Contact ASE, College Lane, Hatfield, Herts. AL10 9AA
  Tel: 01707 267411 Fax: 01707 266532
  http://www.ase.org.uk

- **British Association for the Advancement of Science** (BA) promotes an
  understanding of science and technology and includes a programme for young
  people in the youth section BAYS. BAYS day takes place annually in London in
  March and many regional events are also organised.

Contact: BA, 23 Savile Row, London W1X 2NB
Tel: 0171 9733500
http://www.britassoc.org.uk

- **Chemical Industries Association** (CIA) is a trade association representing the interests of small and large chemical companies. The association actively pursues policies designed to support the teaching of science and technology and encourages local chemical companies to develop education/business partnerships. The CIA together with York University set up the Chemical Industry Education Centre (CIEC).
  Contact: The Chemical Industries Association, King's Buildings, Smith Square, London SW1P 3JJ
  Tel: 0171 8343399 Fax: 0171 8344469

- The **Chemical Industry Education Centre** (CIEC) is dedicated to the effective teaching of science and technology and to a proper level of understanding between schools and the chemical industry. It also offers a specialist research and consultancy service to industry and an independent, comprehensive advice and information service to teachers and schools throughout the UK. CIEC produces a range of publications for schools.
  Contact: CIEC, University of York, Heslington, York YO1 5DD
  Tel: 01904 432523 Fax: 01904 432516

- **Creativity in Science and Technology** (CREST) provide opportunities for students to experience scientific and engineering research at first hand and to couple to this opportunity for national accreditation via Bronze, Silver, Gold and Platinum Awards.
  For further information contact CREST Awards, 1 Giltspur Street, London EC1A 9DD
  Tel: 0171 2942442

- The **Pupil Researcher Initiative** (PRI) is a curriculum development project established to support the teaching and learning of science in the 14–16 age range. It is funded by the Engineering and Physical Sciences Research Council (EPSRC) and the Particle Physics and Astronomy Research Council (PPARC) and is co-ordinated by a team at the Centre for Science and Mathematics Education at Sheffield Hallam University. The PRI provide investigative work and access to PhD placements free to schools.
  Contact: PRI, Centre for Science Education, School of Science and Mathematics, Sheffield Hallam University, Sheffield S10 2BP
  Tel: 01142 532211 Fax: 01142 532211
  e-mail: n.a.fuller@shu.ac.uk
  http://www.shu.ac.uk.schools/sci/pri

- **The Royal Society** is an independent academy promoting the natural and applied sciences, nationally and internationally.
  Contact: The Royal Society, 6 Charlton House Terrace, London SW1 5AG
  Tel: 0171 8395561
  http://www.royalsoc.ac.uk/rs/

- **The Royal Society of Chemistry** (RSC) is the UK's learned society for chemistry and the professional body for chemists. It is concerned with advancing chemistry as a science, disseminating chemical knowledge and developing applications of chemistry. The Education Department produces a wide range of teacher support and careers material. See Bibliography for useful publications.
  Contact: Education Department, The Royal Society of Chemistry, Burlington House, Piccadilly, London
  Tel: 0171 4378656
  http://www.rsc.org.uk

- **Science and Technology Regional Organisation** (SATRO) is the operational arm of the Standing Conference on Schools' Science and Technology (SCSST). There are over 40 SATRO operations delivering the SCSST's national programme throughout the country.
  Contact: Tel: 0171 7535200

Index

A

acid 121
acid rain 111,112
active electrodes 67
addition polymers 87
air 6, 108
alkali metals 103
alkane 83
alkene 83
allotrope 17
alloy 102
amphoteric oxides 111
anion 62
anode 62
anode rules 65
Atlantic ridge 46
atmosphere 108
atom 21, 29
atomic number 21
Avogadro number 78

B

base 121
blast furnace 101
bond 34

bonding 34
bond energies 35

C

carbon dioxide 110
catalysts 54
cathode rules 65
cations 62
ceramics 17
chemical symbols 72
CLISP project 7
collision theory 54
colloid 13
composite 17
compound 21
condensation polymer 87
covalent bond 35
cracking 85
crude oil 81
crystal lattice 7

D

decomposition reaction 51
direct current 61

dissolving 9
displacement reaction 97
dot and cross diagrams 34

E

earth structure 38
earthquakes 46
electron 34
electrodes 60
electrolysis 60
electrolytes 60
electrovalent bond 34
element 21
endothermic reaction 55
energy level diagrams 55
enthalpy 56
epicentre 4.2
equation 46
equilibrium reactions 57
erosion 44
evaporation 17
exothermic reaction 55
extrusive igneous rocks 41

F

fire extinguisher 110
fractional distillation 108
formula 73

G

gas 5, 110
glass 17
graphite electrode 62
graphical formula 84
greenhouse effect 112

groups in the periodic table 117

H

Haber process 112

I

igneous rocks 41
indicator 123
inert electrodes 69
intrusive igneous rocks 41
iodine 16
ionic bond 35
ions 63
isomer 84

K

kinetic theory 6

L

lattice 7
liquid 6
lithosphere 41

M

magma 41
magnetic field 40
mantle 41
mass calculations 79
Mendeleev 111
metals 94
metamorphic rocks 41

mid-Atlantic ridge 45
minerals 41
mixture 21
mole 78
molecules 6
molecular formula 84

N

naming organic molecules 82
neutralisation 122
nitrogen 110
nitrogen oxides 110
nylon 89

O

oil 81
ore 99
organic molecules 82
oxygen 110
ozone depletion 112

P

p waves 40
pH 121
particle 5.6
periodic table 115
periods 115
peridotite 42
photosynthesis 55
plastics 87
plate tectonics 41, 45
polymers 87
polymerisation 87
product 49

R

reactant 49
reaction 49
reactivity series 100
recycling 88
refining oil 81
relative atomic mass 75
respiration 55
rocks 41
rock cycle 44
rusting 101

S

s waves 40
salts 125
saturation 83
sea floor spreading 42
sedimentary rocks 41
solid 5
solvent 12
solution 12
states of matter 5
structural formula 84
subduction 42
sublimation 16
sulphur dioxide 111
symbols 72

T

thermit reaction 98
thermoplastics 92
thermosetting plastics 92

U

unsaturated 83
universal indicator 124

V

valency 74
volcanoes 46

W

weathering 44